Lecture Notes in Earth Sciences 114

Chi-Yuen Wang · Michael Manga

Earthquakes and Water

Springer

Prof.Dr. Chi-Yuen Wang
University of California,
Berkeley
Dept. Earth & Planetary
Science
Berkeley CA 94720-4767
USA
chiyuen@berkeley.edu

Prof. Michael Manga
University of California,
Berkeley
Dept. Earth & Planetary
Science
Berkeley CA 94720-4767
USA
manga@seismo.berkeley.edu

ISSN 0930-0317
ISBN 978-3-642-26324-8 ISBN 978-3-642-00810-8 (eBook)
DOI 10.1007/978-3-642-00810-8
Springer Heidelberg Dordrecht London New York

Cover design: Integra Software Services Pvt. Ltd., Pondicherry

Printed on acid-free paper

Springer is part of Springer Science+Business Media (www.springer.com)

Preface

Distant earthquakes are well known to induce a wide range of responses in surface water and groundwater. These responses are often viewed as mere curiosities as their occurrence is limited in space and time. The frequent emphasis on earthquake precursors in studies of these phenomena also tends to push the study of 'earthquake hydrology' away from the mainstream of geoscience. The observed phenomena, however, probe the interaction between hydrogeological processes and mechanical deformation in the shallow crust. Hence they provide insight into the interaction among water cycle, tectonics, and properties of the crust. As such, the study of earthquake hydrology also has the potential to provide a more quantitative and in-depth understanding of the nature of earthquake precursors and evaluate whether they are in fact precursors.

The title of this book reflects the nature of the connections we address: we focus on how earthquakes affect hydrology. Water also influences earthquakes as it affects the strength of faults and the rheology of rocks. Our emphasis here, however, is not on the hydrology of earthquakes, but on understanding the hydrological phenomena induced or modified by earthquakes. The boundary between the 'hydrology of earthquakes' and the 'earthquake-induced hydrological phenomena', however, can sometimes be blurred. For example, triggered earthquakes are sometimes explained by a re-distribution of pore pressure following the triggering earthquake. Hence, triggered seismicity may be an example of an earthquake-induced hydrological phenomenon. The study of the latter, therefore, can be important towards a better understanding of the mechanics of at least some earthquakes.

There are many students, postdocs and colleagues we wish to thank for collaborating on research projects related to the topics reviewed in this book, or participating in stimulating discussions in the class we taught called 'Earthquake hydrology'. In particular, we wish to thank Emily Brodsky, Yeeping Chia, Douglas Dreger, Shemin Ge, Fu-qiong Huang, Tom Holzer, Chris Huber, Joel Rowland, Martin Saar, Yaolin Shi, Chung-Ho Wang, Kelin Wang, Pei-ling Wang and Alex Wong for enlightening exchanges. Hunter Philson helped with figures and the index. We also thank the National Science Foundation, the Miller Institute for Basic Research in Science, and NASA for supporting the research and synthesis in this volume.

Berkeley, California

Chi-Yuen Wang
Michael Manga

Contents

Chapter 1
Introduction

For thousands of years, a variety of hydrologic changes has been documented following earthquakes. Examples include the liquefaction of sediments, increased stream discharge, changes in groundwater level, changes in the temperature and chemical composition of groundwater, formation of new springs, disappearance of previously active springs, and changes in the activities of mud volcanoes and geysers. It is not unexpected that earthquakes can cause hydrologic changes because the stresses created by earthquakes can be large. What is surprising are the large amplitudes of hydrologic responses and the great distances over which these changes occur. Following the 2004 M9.2 Sumatra earthquake, for example, groundwater erupted in southern China, 3200 km away from the epicenter, and the water fountain shown in Fig. 1.1 reached a height of 50–60 m above the ground surface when it was first sighted. Because earthquakes and water interact with each other through changes in both stress and physical properties of rocks, understanding the origin of hydrological responses can provide unique insight into hydrogeologic and tectonic processes at spatial and temporal scales that otherwise could not be studied.

Earthquakes cause both static and dynamic changes of the stresses in the crust. Both types of stress change decrease with increasing distance from the earthquake, but at different rates. Figure 1.2, from Kilb et al. (2002), illustrates how static and dynamic stress change with increasing distance from the epicenter. The dynamic component of the Coulomb stress change, $\Delta\mathrm{CFS}(t)$, as defined in the caption of Fig. 1.2, is the time-dependent change in the Coulomb failure stress resolved onto a possible failure plane. The static stress change, denoted by $\Delta\mathrm{CFS}$, diminishes much more rapidly with distance than the transient, dynamic change. Thus at close distances the ratio (peak $\Delta\mathrm{CFS}(t))/\Delta\mathrm{CFS}$ is approximately proportional to the source-receiver distance, r, and at larger distances proportional to r^2 (Aki and Richards, 1980). At distances up to ~1 ruptured fault length, the static and the peak dynamic changes are comparable in magnitude, while at distances greater than several ruptured fault lengths, the peak dynamic change is much greater than the static change. As discussed in later chapters, the relative magnitude of the static and dynamic stresses is reflected in the hydrologic responses to earthquakes and is critical to understanding the origin of hydrological changes. We thus hereafter use the expression 'near field' to denote distances within about one ruptured fault

C.-Y. Wang, M. Manga, *Earthquakes and Water*, Lecture Notes in Earth Sciences 114, DOI 10.1007/978-3-642-00810-8_1, © Springer-Verlag Berlin Heidelberg 2010

Fig. 1.1 Well in China responding to the December 26, 2004, M 9.2 Sumatra earthquake 3200 km away. The picture was taken by Hou Banghua, Earthquake Office of Meizhou County, Guangdong, 2 days after the Sumatra earthquake. The fountain was 50–60 m high when it was first sighted 1 day after the earthquake

length, 'far field' to denote distances many times greater than the fault length, and 'intermediate field' for distances in between.

Besides being a matter of academic interest, the study of earthquake-induced hydrologic changes also has important implications for water resources, hydrocarbon exploration and engineering enterprise. For example, groundwater level changes following earthquakes can affect water supplies (Chen and Wang, 2009) and it is sometimes necessary to evaluate the causative role of an earthquake in insurance claims for loss of water supply (Roeloffs, 1998). Furthermore, earthquake-induced increase in crustal permeability (e.g., Rojstaczer et al., 1995; Roeloffs, 1998; Brodsky et al., 2003; Wang et al., 2004; Elkhoury et al., 2006; Wang and Chia, 2008) has important implications on hydrocarbon migration and recovery on the one hand, and contaminant transport on the other. Forensic earthquake hydrology was also applied to evaluate whether an earthquake may have played causative role in the 2006 eruption of a mud volcano eruption in the Indonesian city of Sidoarjo, in eastern Java, that led to massive destruction of property and evacuation of people (Manga, 2007). Groundwater level changes following earthquakes may also put some underground waste repositories at risk (Carrigan et al., 1991; Roeloffs, 1998). Earthquake-induced fluid pressure changes can induce liquefaction of the ground that causes great damage to engineered structures (e.g., Seed and Lee, 1966), affect oil well production (Beresnev and Johnson, 1994), and trigger seismicity (Hill and Prejean, 2007). Finally, measured changes of the pore pressure in rocks and/or

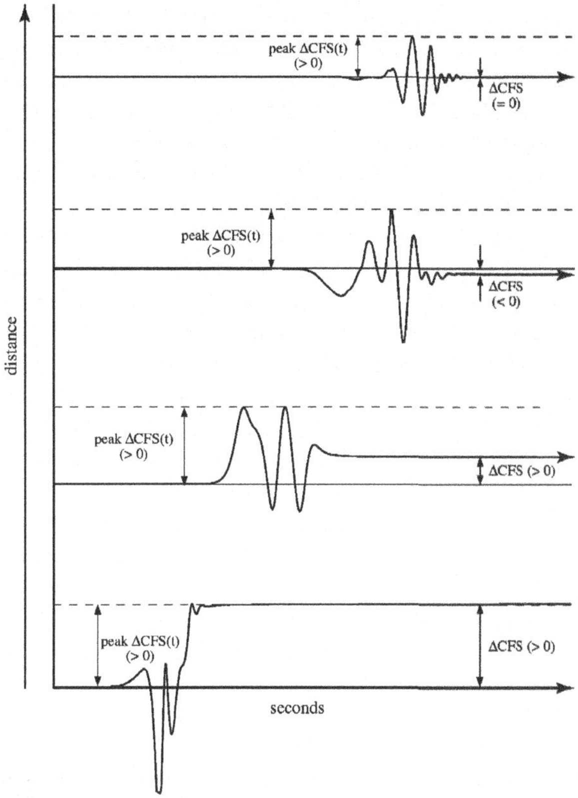

Fig. 1.2 Cartoon illustrating the peak dynamic Coulomb stress change (peak ΔCFS(t)) and static Coulomb stress change (ΔCFS), and their variation with distance from the ruptured fault. $\Delta CFS(t) \equiv \Delta\tau\ (t) - \mu\ [\Delta\sigma_n\ (t) - \Delta P(t)]$, where τ is shear stress on the fault, σ_n is the stress normal to the fault, P is the pore pressure, and μ is the coefficient of friction. In the far field, peak dynamic stresses, ΔCFS(t), are far greater than the static change, ΔCFS, but in the near field, both are comparable in magnitude (Modified from Kilb et al., 2002)

the chemical composition of groundwater are sometimes taken as signatures of the crustal response to tectonic deformation (e.g., Davis et al., 2006) or even as earthquake precursors (e.g., Silver and Wakita, 1996).

In the past decade, there has been a rapid increase in the number and especially the quality of quantitative data documenting hydrological changes during and following earthquakes, largely due to the implementation of hydrological, seismological and geodetic monitoring systems around the globe. Research results on this topic, however, have been published in various journals and various fields (geoscience, hydrology, geotechnical engineering, petroleum geology). We felt it timely and desirable to summarize the advances made so far in a single volume, both in terms of observations as well as their analysis and interpretation. Such a volume may serve on one hand as a convenient reference for researchers active in this area, and, on the other hand, as a starting point for students interested in this topic, and may thus help to advance the studies of earthquake-induced hydrologic changes. A shorter review on this general topic is also available in Manga and Wang (2007).

This volume does not address all possible interactions between water and earthquakes. Not covered are tsunamis, for which useful references may be found in many topical volumes (e.g., Satake et al., 2007). Water plays a critical role in the frictional strength (or weakness) of faults (e.g., Lachenbruch and Sass, 1977; Mount and Suppe, 1987; Zoback et al., 1987; Rice, 1992; Andrew, 2003; Wibberley and Shimamoto, 2005). The presence of a saturated fault gouge zone, with large porosity extending to seismogenic depths was suggested from several lines of studies, including clay mineralogy and texture of fault gouge (Wu, 1978; Rutter, et al., 1986; Schleicher et al., 2009), interpretation of the low seismic wave velocity and low electrical resistivity across the San Andreas fault using laboratory data (Wang et al., 1978; Wang, 1984), analysis of gravity anomalies across the San Andreas fault (Wang et al., 1986), investigations of exhumed fault zones in California (e.g., Chester et al., 2004) and in Japan (Wibberley and Shimamoto, 2005; Forster et al., 2003) and drilled cores from fault zones (e.g., Ohtani et al., 2001), seismic imaging of the fault-zone structures (Mizuno, 2003; Li et al., 2006) and electromagnetic imaging of the San Andreas fault zone (Unsworth et al., 1999). Direct drilling of the San Andreas fault zone to a depth of $\sim$3 km near Parkfield, CA, also revealed a saturated process zone 250 m in width, in which the seismic velocities are 20–30% below the wall-rock velocities (Hickman et al., 2005). These topics, while beyond the objective of this book, highlight the importance of understanding better the interaction between water and earthquakes.

In Chap. 2–8 we discuss separately several types of hydrologic responses to earthquakes, i.e., liquefaction, mud volcanoes, increased streamflow, groundwater level changes, changes in groundwater temperature and composition, changes in geyser activities, and triggered earthquakes. In Chap. 9, we summarize the current state of understanding on hydrologic precursors before earthquakes. The concepts of dynamic strain and seismic energy density are used interchangeably throughout the book. The latter is defined in Chap. 2 as the maximum seismic energy available to do work in a unit volume and is easily estimated from the earthquake magnitude and the distance from the earthquake source (Chap. 2). It thus provides a convenient metric to relate and compare the different hydrologic responses. Using this metric, we integrate and compare the various hydrologic responses in the last chapter (Chap. 10) and provide a coherent explanation to all these responses.

The basic theories and equations for groundwater flow, groundwater transport and hydro-mechanical coupling needed to understand some material in the book are provided in the appendices. Also included in the appendices are the definitions of the notation adopted. An extensive listing of the hydrologic data and their sources used in this volume in developing the hypotheses is provided at the end of the Appendices for readers to assess the reliability of the source material, to check the models developed in this volume, or to develop and test their own hypotheses.

References

Aki, K., and P.G. Richards, 1980, *Quantitative Seismology*, New York: W.H. Freeman and Company.

Andrew, D.J., 2003, A fault constitutive relation accounting for thermal pressurization of pore fluid, *J. Geophys. Res., 107*, doi:10.1029/2002JB001942.

Beresnev, I.A., and P.A. Johnson, 1994, Elastic wave stimulation of oil production: A review of methods and results, *Geophys., 59*, 1000–1017.

Brodsky, E.E., E. Roeloffs, D. Woodcock, I. Gall, and M. Manga, 2003, A mechanism for sustained groundwater pressure changes induced by distant earthquakes, *J. Geophys. Res., 108*, 2390, doi: 10.1029/2002JB002321.

Carrigan, C.R., G.C.P. King, G.E. Barr, and N.E. Bixler, 1991, Potential for water-table excursions induced by seismic events at Yucca Mountain, Nevada, *Geology, 19*, 1157–1160.

Chen, J.S., and C.-Y. Wang, 2009, Rising springs along the Silk Road, *Geology, 37*, 243–246.

Chester, F.M., J.S. Chester, D.L. Kirschner, S.E. Schulz, and J.P. Evans, 2004, Structure of large-displacement strikeslip fault zones in the brittle continental crust. In: G.D. Karner, B. Taylor, N.W. Driscoll, and D.L. Kohlstedt (eds.), *Rheology and Deformation in the Lithosphere at Continental Margins*, pp. 223–260. New York: Columbia University Press.

Davis, E.E., K. Becker, K. Wang, K. Obara, Y. Ito, and M. Kinoshita, 2006, A discrete episode of seismic and aseismic deformation of the Nankai trough subduction zone accretionary prism and incoming Philippine Sea plate, *Earth Planet. Sci. Lett., 242*, 73–84.

Elkhoury, J.E., E.E. Brodsky, and D.C. Agnew, 2006, Seismic waves increase permeability, *Nature, 411*, 1135–1138.

Forster, C.B., J.P. Evans, H. Tanaka, R. Jeffreys, and T. Noraha, 2003, Hydrologic properties and structure of the Mozumi Fault, central Japan, *J. Geophys. Res., 30*, doi:10.1029/2002GL014904.

Hickman, S.H., M.D. Zoback, and W.L. Ellsworth, 2005, Structure and composition of the San Andreas fault zone at Parkfield: Initial results from SAFOD, phase 1 and 2. *EOS Trans., 83*, 237, American Geophysical Union.

Hill, D.P., and S.G. Prejean, 2007. Dynamic triggering. In: *Treatise on Geophysics*, G. Schubert editor, Vol. 4, pp. 257–292.

Kilb, D., J. Gomberg, and P. Bodin, 2002, Aftershock triggering by complete Coulomb stress changes, *J. Geophys. Res., 107*(B4), 2060, doi:10.1029/2001JB000202.

Lachenbruch, A.H., and J.H. Sass, 1977, Heat flow in the United States and the thermal regime of the crust. In: J.H. Heacock (ed.), *The Earth's Crust, Geophys. Monogr. Ser., 20*, pp. 625–675. Washington, DC: American Geophysical Union.

Li, Y.G., J.E. Vidale, and P.E. Malin, 2006, Parkfield fault-zone guided waves: High-resolution delineation of the low-velocity damage zone on the San Andreas at depth near SAFOD site, *Proc. ICDP-IODP Fault-Zone Drilling*, pp. 1–4. Miyazaki, Japan.

Manga, M., 2007, Did an earthquake trigger the may 2006 eruption of the Lusi mud volcano? *EOS, 88*, 201.

Manga, M. and C.-Y. Wang, 2007, Earthquake hydrology, H. Kanamori (ed.), *Treatise on Geophysics, 4*, Elsevier, Ch. In: *Treatise on Geophysics*, G. Schubert editor, Vol. 4, pp. 293–320.

Mizuno, T., 2003, The subsurface observation of fault-zone trapped waves: Applications to investigations of the deep structure of active faults. *Bull. Earthq. Res. Inst., 78*, 91–106.

Mount, V.S., and J. Suppe, 1987, State of stress near the San Andreas fault – implications for wrench tectonics, *Geology, 15*, 1143–1146.

Ohtani, T., H. Tanaka, K. Fujimoto, T. Higuchi, N. Tomida, and H. Ito, 2001, Internal structure of the Nojima fault zone from Hirabayshi GSJ drill core, *Island Arc, 10*, 392–400.

Rice, J.R., 1992, Fault stress state, pore pressure distributions, and the weakness of the san Andreas fault. In: B. Evans, and T.-F. Wong (eds.), *Fault Mechanics and Transport Properties of Rocks*, pp. 475–504. San Diego: Academic Press.

Roeloffs, E.A., 1998, Persistent water level changes in a well near Parkfield, California, due to local and distant earthquakes, *J. Geophys. Res., 103*, 869–889.

Rojstaczer, S., S., Wolf, and R., Michel, 1995, Permeability enhancement in the shallow crust as a cause of earthquake-induced hydrological changes, *Nature, 373*, 237–239.

Rutter E.H., R.H. Maddock, S.H. Hall, and S.H. White, 1986, Comparative microstructure of natural and experimentally produced clay bearing fault gouges, *Pure Appl. Geophys., 24*, 3–30.

Satake, K., E.A. Okal, and J.C. Borrero, 2007, *Tsunami and Hazards in the Indian and Pacific Oceans*, 392 p, Basel: Birkhauser.

Schleicher, A.M., L.N. Warr, and B.A. van der Pluijm, 2009, On the origin of mixed-layered clay minerals from the San Andreas Fault at 2.5–3 km vertical depth (SAFOD drillhole at Parkfield, California), *Contrib. Mineral. Petrol., 157*, 173–187.

Seed, H.B., and K.L. Lee, 1966, Liquefaction of saturated sands during cyclic loading, *J. Soil Mech. Found. Div., 92*, 105–134.

Silver, P.G., and H. Wakita, 1996, A search for earthquake precursors, *Science, 273*, 77–78.

Unsworth, M.J., G. Egbert, and J. Booker, 1999, High-resolution electromagnetic imaging of the San Andreas Fault in Central California, *J. Geophys. Res., 104*, 1131–1150.

Wang, C.-Y., W. Lin, and F. Wu, 1978, Constitution of the San Andreas fault zone at depth, *Geophys. Res. Lett., 5*, 741–744.

Wang, C.-Y., 1984, On the constitution of the San Andreas fault, *J. Geophys. Res., 89*, 5858–5866.

Wang, C.-Y., F. Rui, Z. Yao, and X. Shi, 1986, Gravity anomaly and density structure of the San Andreas fault zone, *Pure App. Geophys., 124*, 127–140.

Wang, C.-Y., and Y. Chia, 2008, Mechanism of water level changes during earthquakes: Near field versus intermediate field, *Geophys. Res. Lett., 35*, L12402, doi:10.1029/2008GL034227.

Wang, C.-Y., C.H. Wang, and M. Manga, 2004, Coseismic release of water from mountains: Evidence from the 1999 (M_w=7.5) Chi-Chi earthquake, *Geology, 32*, 769–772.

Wibberley, C., and T. Shimamoto, 2005, Earthquake slip weakening and asperities explained by thermal pressurization, *Nature, 436*, 689–692.

Wu, F.T., 1978, Mineralogy and physical nature of clay gouge, *Pure Appl. Geophys., 116*, 655–689.

Zoback, M.D., M.L. Zoback, V.S. Mount, et al., 1987, New evidence on the state of stress of the San Andreas fault system, *Science, 238*, 1105–1111.

Chapter 2
Liquefaction

Contents

2.1 Introduction

In 373/2 BC Helice, a coastal town in ancient Greece, disappeared entirely under the sea after being leveled by a great earthquake. In 1861, the same place was hit by another earthquake, though to a lesser degree. Schmidt (1875) studied the affected area and showed extensive lateral spreading and subsidence of land along the coast (Fig. 2.1). Similar phenomena were documented along the coast near Anchorage, Seward and Valdez following the 1964 M9.2 Alaska earthquake (see Sect. 2.2), where the slumping of land was shown to be caused by the liquefaction of soft clays and sands underneath a gentle slope (Seed, 1968).

Liquefaction is a process by which the rigidity of saturated sediments is reduced to zero and the sediments become fluid-like. It occurs mostly during earthquakes and is invariably associated with high pore-water pressure, as evidenced by the common

C.-Y. Wang, M. Manga, *Earthquakes and Water*, Lecture Notes in Earth Sciences 114, DOI 10.1007/978-3-642-00810-8_2,

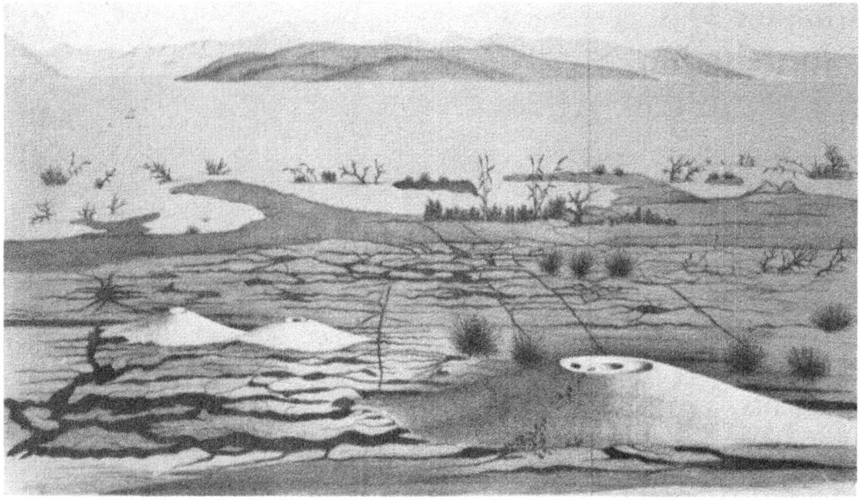

Fig. 2.1 Drawing shows the affected area of Helice after the earthquake of 1861. In the foreground, the remaining part of the land was broken into a collage of many irregular pieces separated by a patchwork of extensional fractures, covered sparingly by sand-craters. Off the coast in the Gulf of Corinth, tree tops marked part of the submerged strip of coastal plain. (From Marinatos, 1960)

occurrence of ejections of water and sediments to substantial height during liquefaction (Fig. 2.2). Eyewitnesses to the great 1964 Alaska earthquake at distances up to 400 km from the epicenter, for example, reported that eruptions of water and sediments reached heights up to 30 m (Waller, 1966).

In addition to being a significant hydrogeologic process in its own right, liquefaction has drawn much attention from engineers because it can create great damage to man-made structures. It causes ground to subside and to spread laterally, thus induces buildings to tilt, damages airport runways and earth embankments, and disrupts buried pipes and pile foundations. Since the 1960s, earthquake engineers have carried out a great amount of research to study liquefaction and to predict its occurrence. Their works are summarized in several special volumes (e.g., National Research Council, 1985; Pitilakis, 2007) and will not be repeated here. Only the results critical to the understanding of the interaction between earthquakes and water are summarized in Sect. 2.3.

The engineering approach to study liquefaction has been based on the principle of effective stress (Appendix D, Eq. D.1 and D.2), first proposed by Terzaghi (1925). Based on this concept, liquefaction is a consequence of pore-pressure increase when sediments consolidate in an 'undrained' condition during earthquakes (see Sect. 2.3). When pore pressure becomes so high that the effectives stress is reduced to zero, sediments become fluid-like, i.e., liquefy. Recent investigation (Wang, 2007) shows, however, that 'undrained' consolidation of sediments may occur only in the near field of an earthquake; beyond the near field the seismic energy density may be too small to initiate consolidation, even in the most sensitive sediments (Sect. 2.4.1).

Fig. 2.2 Smear *left* on a building wall created by ejected sand during the 1999 M 7.5 Chi-Chi earthquake in Taiwan. Motorcycle at lower *left* shows scale (From Su et al., 2000)

A new mechanism may thus be required to explain the great number of liquefaction examples that have been documented beyond the near field (Fig. 2.6; Sect. 2.4.2).

In Sect. 2.5, we discuss the dependence of liquefaction on the frequency of the seismic waves. This issue is important but has been scarcely studied until recently. We emphasize that the results remain controversial and require more research in the future.

2.2 Observations in the Near Field

One of the best studied regions for liquefaction features is the New Madrid Seismic Zone in the central United States (Fig. 2.3), where widespread liquefaction was induced by nearby historic and prehistoric earthquakes. Liquefaction features, mapped over several thousand square kilometers (Obermeir, 1989), are present in various shapes, sizes, and ages. Many surficial vented deposits, or sand blows, are 1.0–1.5 m in thickness and 10–30 m in diameter and are still easy to identify on the ground surface and on aerial photographs and satellite images despite years of modification by active agricultural activities (Tuttle and Schweig, 1996). Sand dikes, which represent the conduits for escaping pore water and sediments from the liquefied layers below the sand blows, are also abundant. Most of these features are thought to have formed during the 1810–1811 M8 New Madrid earthquakes, even though many may be prehistoric in age (Tuttle and Schweig, 1996).

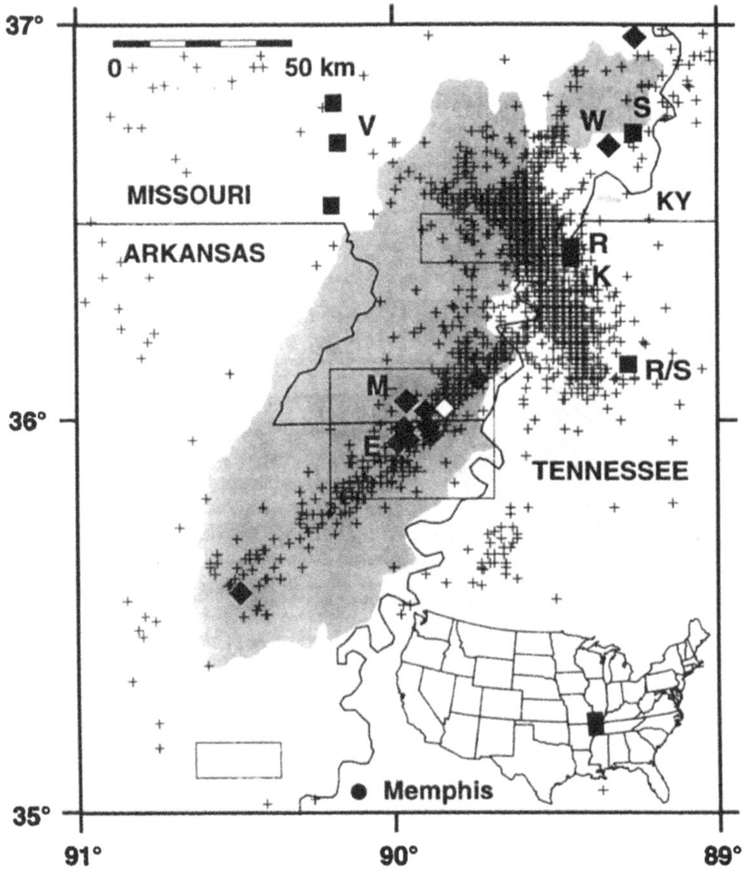

Fig. 2.3 Location map of showing the liquefaction sites within the New Madrid seismic zone. *Shaded area* represents the area where >1% of the ground surface is covered by sand-blow deposits (Obermeir, 1989). Seismicity (1974–1991), shown by *crosses*, defines the New Madrid Seismic Zone. *Symbols* and *letters* refer to sites of previous liquefaction and paleoliquefaction studies (From Tuttle and Schweig, 1996)

Two earthquake events are particularly important in bringing liquefaction phenomena and their devastating effects to the attention of engineers and seismologists. This awareness has in turn led to a great amount of research during the past 50 years in an effort to better understand liquefaction and to mitigate its damage.

The 1964 M9.2 Alaska earthquake occurred at a depth of approximately 30 km beneath Prince William Sound; the rupture extended laterally for 800 km parallel to the Aleutian trench and uplifted about 520,000 km^2 of the crust. Many landslides occurred; the most spectacular slide took place at the Turnagain Height area of Anchorage, caused by liquefaction of the underlying soft clay and sands. The slide extended ~2800 m laterally along a bluff and continued inland for an average distance of ~300 m, resulting in 130 acres of land sliding toward the ocean (Seed, 1968). Within the slide area the ground was broken into blocks that collapsed and tilted at all angles forming a chaotic collage of ridges and depressions.

Fig. 2.4 Tilted apartment buildings after the Nigata earthquake. Despite the extreme tilting, the building themselves suffered remarkably little structural damage (From the Earthquake Engineering Research Center Library, University of California at Berkeley)

In the depressed areas, the ground dropped an average of 12 m during the sliding. Houses in the area, some of which moved laterally as much as 150 or 180 m, were completely destroyed.

During the 1964 M7.5 Nigata Earthquake, Japan, dramatic damage was caused by liquefaction of the sand deposits in the low-lying areas of Nigata City. The soils in and around this city consist of recently reclaimed land and young sedimentary deposits having low density and a shallow ground water table. About 2,000 houses in Nigata City were totally destroyed; more than 200 reinforced concrete buildings tilted rigidly without appreciable damage to the structure (Fig. 2.4).

In most cases the liquefied sediments are sand or silty sand. However, well-graded gravel has increasingly been witnessed to liquefy during recent earthquakes. During the 1983 Borah Peak earthquake in Idaho, for example, fluvial sandy gravel liquefied extensively (Youd et al., 1985). Another example is the extensive liquefaction of reclaimed land in Kobe during the 1995 Hyogoken Nambu earthquake in Japan (Kokusho, 2007), which was filled with gravel-sized granules and fines of decomposed granite.

2.3 Laboratory Studies

Terzaghi's principle of effective stress (see Appendix 3), first proposed in the early twentieth century (Terzaghi, 1925), laid the foundation for soil mechanics and later for earthquake engineering. The structural integrity of sediments, which allows the sediments to carry weight, is normally maintained through grain-to-grain contacts. Seismic shaking can disturb the grain-to-grain contacts to cause sediments to consolidate. Some weight initially carried by the sediments is then shifted to the interstitial pore water. Since the duration of seismic shaking, normally tens of seconds, is short compared to the time required to dissipate pore pressure in the sediment, consolidation of saturated sediments occurs in an 'undrained' condition, and pore pressure builds up. As a result, the 'effective stress' supported by the sediments decreases correspondingly. If pore pressure continues to increase and the effective stress vanishes, all the weight-carrying capacity of the sediments is lost and the weight of the sediments is born entirely by the pore water and the sediments become fluid-like.

Based on the principle of effective stress, earthquake engineers have carried out a great many laboratory experiments in the past half a century to better understand the processes of undrained consolidation of saturated sediments under cyclic loading, the ensuing pore-pressure buildup, and the eventual occurrence of liquefaction. Figure 2.5 shows the changes in shear stress and pore pressure in a sand specimen under cyclic shearing at constant strain amplitude of $\pm 2 \times 10^{-3}$. Pore pressure

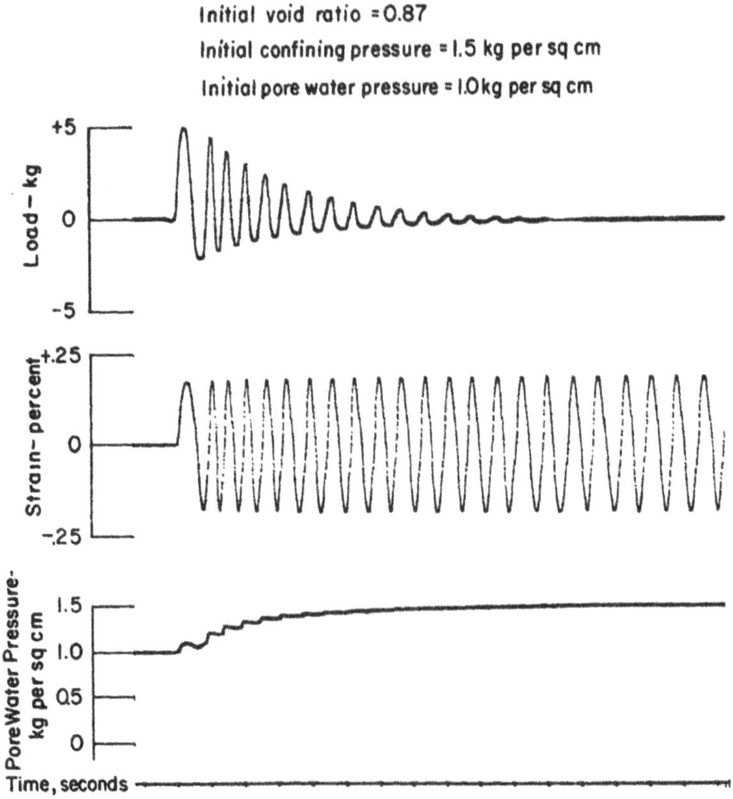

Fig. 2.5 Pore-pressure generation and axial stress in a saturated sand specimen subjected to cyclic shearing at constant strain of $\pm 2 \times 10^{-3}$. The sediment specimen had an initial dry density of 87%. The sample was subjected to a confining pressure of 15 KPa and an initial pore pressure of 10 KPa. Note that the axial stress declined with increasing number of cycles, showing a continued weakening. The rate of pore pressure increase also declined with increasing number of cycles (From Seed and Lee, 1966)

increased gradually from the beginning of shearing until it reaches the magnitude of the confining pressure. With the increasing pore pressure, the sample weakens as demonstrated by the diminishing shear stress it supports. The sample fluidizes when its shear strength approaches zero and the pore pressure approaches the magnitude of the confining pressure. These changes are complimentary to those obtained in the constant stress experiment (Figs. D.4 and D.5).

In the following Sect. 2.3.1 we review the important developments in cyclic loading experiments designed to understand the processes of liquefaction under seismic shaking. In the next subsection we summarize the experimental results on the dissipated energy required to initiate liquefaction by undrained consolidation. This provides the basis for the discussion in Sect. 2.4 in which we show that undrained

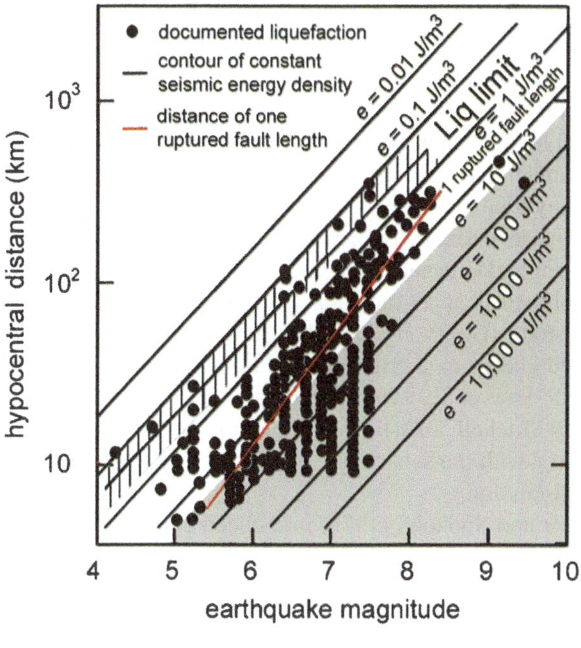

Fig. 2.6 Global dataset for hypocentral distance of documented liquefactions, shown in *solid circles*, is plotted against earthquake magnitude; hachured band marks the threshold distance of liquefaction (i.e., the liquefaction limit) as a function of earthquake magnitude (Wang et al., 2006). *Black lines* are contours of constant seismic energy densities (Eq. 2.4). *Grey line* is the empirical relation between the epicentral distance of 1 ruptured fault length and earthquake magnitude (see text for explanation), which defines the outer boundary of the 'near field'. The upper boundary of the *shaded area* corresponds to a contour with $e = e_u = 30$ J/m^3 – the minimum dissipated energy density required to initiate consolidation-induced liquefaction in sensitive sediments (see text for explanation); thus the *shaded area* is associated with seismic energy densities greater than 30 J/m^3. Note that the shaded region lies mostly in the near field; note also that abundant liquefactions occurred at distances beyond the near field up to distances of several ruptured fault lengths, i.e., the 'intermediate field', with the liquefaction limit located where the seismic energy density falls in a range between 0.1 and 1 J/m^3 (From Wang, 2007)

consolidation may occur only in the near field of an earthquake and that, beyond the near field, the seismic energy density may be too small to induce undrained consolidation. On the other hand, a great number of liquefaction features have been documented beyond the near field (Fig. 2.6; Wang, 2007). Thus, while Terzaghi's principle may explain the occurrence of liquefaction within the near field, a new mechanism is required to explain the occurrence of liquefaction beyond the near field.

2.3.1 Cyclic Loading Experiments

Much of the current understanding of the mechanism of liquefaction is based on the results of a large number of laboratory experiments in geotechnical engineering, in which saturated sediments are subjected to cyclic loading (e.g., National Research Council, 1985). The major objective of these studies is to quantify how pore pressure changes in the cyclically loaded sediments and how sediments eventually liquefy. Despite the different experimental designs, i.e., cyclic torsional shearing of cylinders in a triaxial loading apparatus (e.g., Liang et al., 1995) and special shake tables designed to operate in large centrifuge machines (e.g., Dief, 2000), the sediment samples are all hydraulically isolated from their surroundings, i.e., the experiments are conducted in an undrained condition. The results of these experiments have been variously applied to evaluate the liquefaction potential of sediment sites, either using a threshold stress as an indicator (e.g., Seed and Idriss, 1967; Youd, 1972), or a threshold strain (Dobry et al., 1982; Vucetic, 1994; Hsu and Vucetic, 2004), or the dissipated energy as a criterion (Nemat-Nasser and Shokooh, 1979; Berrill and Davis, 1985; Law et al., 1990; Figueroa et al., 1994; Liang et al., 1995; Dief, 2000; Green and Mitchell, 2004). The last of these, i.e., the dissipated energy, can be directly compared with the seismic energy in the field and is most directly relevant to the present discussion.

Nemat-Nasser and Shokooh (1979) introduced the concept of dissipated energy for the analysis of densification and liquefaction of sediments. Berrill and Davis (1985), Law et al. (1990) and Figueroa et al. (1994) established relations between pore pressure development and the dissipated energy during cyclic loading to explore the use of energy density in the evaluation of the liquefaction potential of sediments. Liang et al. (1995) conducted torsional triaxial experiments on hollowed cylinders of sand to examine the effect of relative density, initial confining pressure and shear-strain magnitude and determined the energy per unit volume (i.e., dissipated energy density) accumulated up to liquefaction; they showed that the dissipated energy density required to induce liquefaction is a function of the relative density of the sediment and the confining pressure. Dief (2000) carried out shake table experiments in a centrifuge with scaled models under a wide range of physical conditions. Dief (2000) also determined the energy density accumulated up to the point of liquefaction and compared the results with those of earlier studies.

2.3.2 *Dissipated Energy for Liquefaction by Undrained Consolidation*

Given the experimental time-histories of shear stress and strain (e.g., Fig. D.5), the cumulative dissipated energy density required to initiate liquefaction by undrained consolidation may be determined by performing the following integral (Berrill and Davis, 1985):

$$e_u = \int_0^t \tau d\gamma \tag{2.1}$$

where τ is the shear stress, γ the shear strain, and the integration extends from the beginning of the cyclic loading to the onset of liquefaction. For the most part, the stress-strain relation varies with each loading cycle, thus the integral can only be evaluated by numerical integration of the experimental stress and strain time histories.

Through such integration, Liang et al. (1995) estimated a dissipated energy density for liquefaction ranging from 290 to 2,700 J/m^3 for sediments with relative densities ranging from 51 to 71% subjected to confining pressures ranging from 41 to 124 KPa; Dief (2000) estimated a dissipated energy density ranging from 470 to 1,700 J/m^3 for relative densities ranging from 50 to 75% subjected to an equivalent confining pressures of ~30 KPa; and Green and Mitchell (2004) obtained a dissipated energy density ranging from 30 to 192 J/m^3 for *clean sand* at an effective confining pressure of 100 KPa. Thus there is a wide range in the dissipated energy density required to induce liquefaction for the ranges of sediment type, relative density and confining pressure studied. The large discrepancies among the different studies may be expected in view that sediments vary widely in their hydromechanical properties and the wide range of experimental conditions. Assuming that the sediment types, the relative density, and confining pressures in these studies are representative for the field conditions relevant to liquefaction, we may take the low value 30 J/m^3, as determined by Green and Mitchell (2004) for clean sand, as the lower bound for the dissipated energy density required to induce liquefaction in the field. The lower bound imposes a threshold seismic energy density required to initiate *consolidation-induced* liquefaction in the field, which, in turn, sets a maximum distance from the earthquake source, as shown in the next section, beyond which consolidation-induced liquefaction may not be expected. The maximum distance so estimated may then be compared with the actual occurrence of liquefaction in the field to verify the hypothesis of undrained consolidation.

2.4 Liquefaction Beyond the Near Field

In Fig. 2.6 we show a recent compilation of global data for liquefaction (Wang, 2007), in which the hypocentral distance r of the documented liquefaction site is plotted against earthquake magnitude M. These parameters, i.e., r and M, are used to characterize the liquefaction occurrences because the majority of documentations (many historical) do not note the style of faulting, the directivity of fault rupture, or the distance to the ruptured fault, nor do they make a distinction among the different magnitude scales. As shown by several authors (Kuribayashi and Tatsuoka, 1975; Ambraseys, 1988; Papadopoulos and Lefkopulos, 1993; Galli, 2000; Wang et al., 2006; Wang, 2007), the occurrence of liquefaction at a given M is delimited by a maximum distance – the liquefaction limit. Since the susceptibility of sediments to liquefaction varies significantly with sediment type and grain size (Seed and Lee, 1966; National Research Council, 1985; Dobry et al., 1982; Hsu and Vucetic, 2004), sediments that liquefy at the liquefaction limit are likely those with the least resistance.

Also shown in Fig. 2.6 is an empirical relation between M and the epicentral distance equal to 1 ruptured fault length for all subsurface fault types (Wells and Coppersmith, 1994). While a large number of the documented liquefactions occur in the near field, i.e., at epicentral distances less than or equal to one ruptured fault length, an equally large number occurred beyond the near field at distances up to several ruptured fault lengths.

2.4.1 *Seismic Energy Density as a Metric for Liquefaction Distribution*

The seismic energy density e at a site during ground shaking may be estimated from the time histories of particle velocity v of the ground motion as recorded by strong-motion seismometers (Lay and Wallace, 1995):

$$e = \frac{1}{2}\sum_i \frac{\rho}{T_i} \int v_i(t)^2 \, dt, \tag{2.2}$$

where the summation is taken over all the relevant modes of the ground vibrations, ρ is density, and T_i and v_i are, respectively, the period and the velocity of the ith mode. Since most energy in the ground motion resides in the peak ground velocity, PGV, we may simplify the above relation to

$$e \sim PGV^2. \tag{2.3}$$

This relation was shown to be consistent with field data in Wang et al. (2006).

Using ~ 30,000 strong-motion records for southern California earthquakes, Cua (2004) showed that the peak ground velocity attenuates with the epicentral distance as $\sim 1/r^{1.5}$ for sediment sites. It follows from (2.3) that,

$$e(r) \cong A/r^3, \tag{2.4}$$

where A is an empirical parameter for southern California. Note that this relation can only be taken as a first-order approximation (i.e., a point-source approximation for the earthquake) because the effect of source dimension and rupture directivity, which become important in determining the distribution of seismic energy in the near field, are not included. Given this approximation, the total seismic energy of an earthquake, E, is related to the energy density at $r = 1$ m, thus to A, by

$$E = \frac{4\pi}{3} e\,(r = 1\ \mathrm{m}) \cong \frac{4\pi}{3} A \tag{2.5}$$

Hence $A \cong 3E/4\pi$; inserting this into (2.4) we have:

$$e\,(r) \cong \frac{3E}{4\pi} r^{-3}. \tag{2.6}$$

Note that this relation is entirely empirical and includes both the geometrical and physical dissipation of the seismic energy. Because of its empirical nature, this relation is strictly valid only for southern California, and may show significant differences from region to region. But, for the reason that no similar relation is yet available elsewhere, we will take it as generally valid.

Combining the above relation with Bath's (1966) empirical relation between E and earthquake magnitude M, we obtain the following empirical relation among e, r and M (Wang, 2007):

$$\log r = 0.48\,M - 0.33 \log e\,(r) - 1.4 \tag{2.7}$$

where r is in km. This relation is plotted in Fig. 2.6 as straight lines for different values of e.

Studies show that the threshold strain required to initiate undrained consolidation in the field is the same as that in the laboratory (Hazirbaba and Rathje, 2004). Thus it may be justified to compare the seismic energy density in the field with the laboratory-based dissipated energy required to initiate liquefaction. Given the discussion in the last section on the laboratory-determined dissipated energy required to initiate liquefaction, we may tie the *maximum* distance of liquefaction occurrence due to undrained consolidation with the contour of $e = e_u = 30$ J/m^3, shown by the upper boundary of the shaded area in Fig. 2.6. As the figure shows, this *maximum* distance corresponds closely with the hypocenter distance of ~1 ruptured fault length, i.e., the outer boundary of the near field. Thus undrained consolidation may account for liquefaction only in the near field. However, as noted earlier, abundant liquefaction are documented at distances far beyond the near field (Fig. 2.6), where the seismic energy density is much below the threshold energy required to induce undrained consolidation. At the maximum distance of liquefaction occurrence, i.e., the liquefaction limit, the seismic energy density declines to ~0.1 J/m^3 which is two orders of magnitude smaller than the threshold energy required to induce undrained consolidation.

2.4.2 Mechanism for Liquefaction Beyond the Near Field

Since the seismic energy density beyond the near field is smaller than the threshold energy density required to initiate consolidation, a different mechanism is needed to account for the occurrence of liquefaction in the intermediate field. An important point to note is that, even though the seismic energy density in the intermediate field is not large enough to induce sediment liquefaction by undrained consolidation, it nonetheless may move the sediments towards a critical state so that they may liquefy if an additional increment of pore pressure becomes available to push the sediments over the liquefaction limit. In the following, we examine whether the spreading of pore pressure from a source to surrounding sediments may be a viable mechanism to provide this additional pore pressure needed to produce liquefaction.

As discussed in Chap. 5, changes of the water level in wells have long been reported after earthquakes and taken to indicate a change in the pore pressure in the groundwater system. Persuasive evidence has become available recently to show that earthquakes can enhance permeability to allow a spread of pore pressure from one part of the hydraulic system to another. Figure 2.7, for example, shows the water-level records from closely spaced wells which monitored the water levels in vertically separated aquifers in central Taiwan before and after the 1999 Chi-Chi earthquake (Water Resource Bureau, 1999). Before the earthquake, the water levels in these wells were distinct, showing that the different aquifers were hydraulically

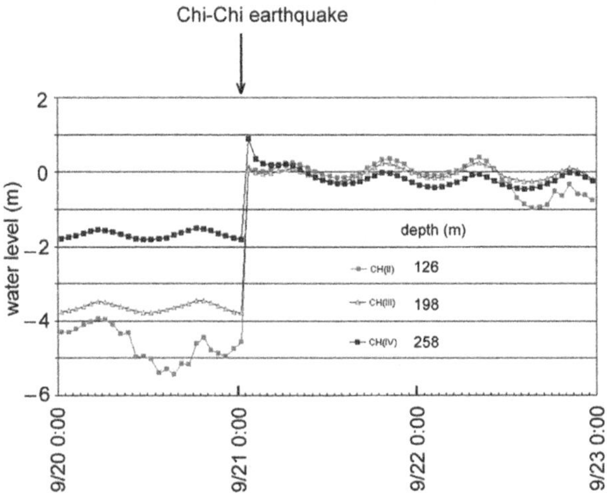

Fig. 2.7 Water-levels in three confined aquifers at a hydrological station (Chuanhsin) in central Taiwan before and after the Chi-Chi earthquake (Water Conservancy Agency, 1999). Hourly data are displayed; different symbols show data for different aquifers. Fine ticks on the horizontal axis are hourly markings; the depths of the aquifers are also given. The sinusoidal oscillations in water levels correspond to the semidiurnal tides. The water levels in the three aquifers were distinct before the earthquake, but became nearly identical after the earthquake, suggesting the aquifers were hydraulically connected by enhanced vertical permeability during the earthquake (From Wang, 2007)

isolated from each other; following the earthquake, however, the water-levels in all these wells came to the same level, suggesting that there was an enhanced *vertical* permeability that connected the different aquifers during the earthquake. The above observation was made in the near field of the earthquake. In the intermediate field, Elkhoury et al. (2006) used the tidal response of water level in wells to measure permeability over a 20-year period and showed distinct transient shifts in the phase of the water level in response to earthquakes at distances beyond the near field. They interpreted these phase shifts in terms of an enhanced permeability induced by earthquakes and attributed the increase in permeability to the removal of colloidal particles from clogged fractures by the seismic waves (Brodsky et al., 2003).

As argued above, the occurrence of consolidation-induced liquefaction may be limited to the near field of an earthquake. To explain the occurrence of liquefaction beyond the near field, we invoke the mechanism of pore-pressure spreading from nearby sources to sediment sites. Since pore-pressure heterogeneity may be the norm in the field, an enhancement of permeability among sites of different pore pressures may cause pore pressure to spread (Roeloffs, 1998; Brodsky et al., 2003; Wang, 2007). Such processes may provide an additional increment of pore pressure to push some critically stressed sediments over the critical state to become liquefied. A corollary to this hypothesis is that during the evaluation of the liquefaction potential of a site, it may be important to consider the hydrogeologic environment of the site, in addition to evaluating the liquefaction susceptibility of the site in isolation.

Finally, we note that the seismic energy density for liquefaction at the threshold distance is minute ($\sim$0.1 J/m^3). What is the mechanism(s) that may cause permeability to increase at such small seismic energies? The model proposed by Brodsky et al. (2003), that the removal of colloidal particles from clogged fractures may enhance the permeability of fractured rocks, implies that a minute amount of seismic energy may suffice to initiate a redistribution of pore pressure. Direct laboratory measurements, however, are needed to quantify the process in order to test this hypothesis. More detailed discussion on the mechanism of removal of colloidal particles from clogged fractures and pores is given in Chap. 5 (Sect. 5.3.3).

2.5 Experiment at Wildlife Reserve, California

The Wildlife liquefaction array was a field experimental array established in 1982 on a flood plain in southern California, about 10 km southeast of the Salton Sea (Fig. 2.8a), and designed specifically to study liquefaction processes. The array (Fig. 2.8b) consisted of two 3-component accelerometers placed at the surface and in a cased borehole at a depth of $\sim$7 m, and six pore-pressure transducers placed around the accelerometers at various depths up to 12 m. Both the M6.2 Elmore earthquake and the M6.6 Superstition Hills earthquake triggered the accelerometers, but only the latter earthquake triggered liquefaction at the array, which caused sand boils with eruptions of water and sediments. Extensive ground cracking implied lateral spreading at the array (Holzer et al., 1989).

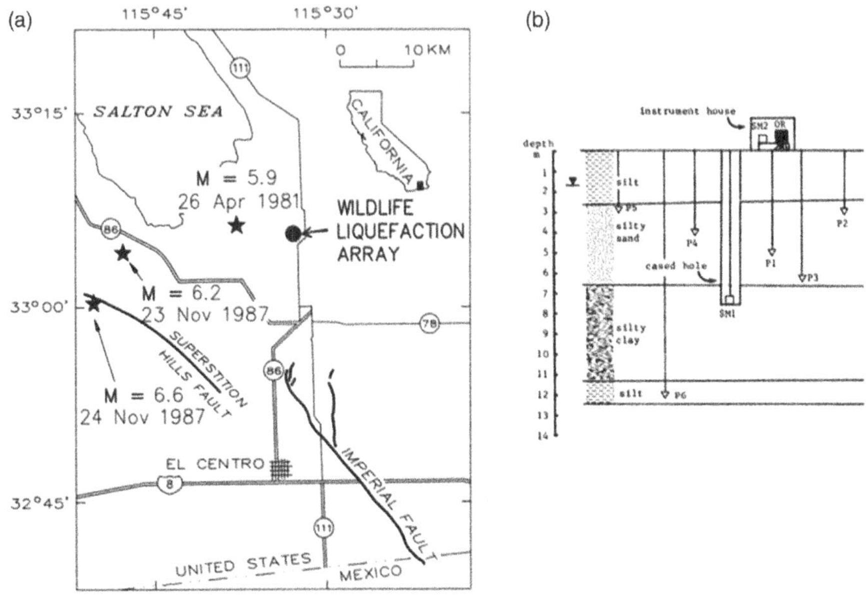

Fig. 2.8 (**a**) Location map of the Wildlife Reserve Array and earthquake epicenters. M6.6 is the 1987 Superstition Hills earthquake, M6.2 is the 1987 Elmore Ranch earthquake, and M5.9 is the 1981 Westmorland earthquake. (**b**) Stratigraphic cross-section of array and schematic of instrument deployment. In plain view, pore-pressure transducers (denoted by P) are equally spaced on the perimeter of a circle with a diameter of 9.1 m. Accelerometers are near center of circle (From Holzer et al., 1989)

The in situ time histories of pore pressure and acceleration at Wildlife Reserve Array during and following the Superstition Hills earthquake reveal a complex interaction among ground shaking, pore pressure buildup and liquefaction (Fig. 2.9). For the convenience of description, Zeghal and Elgamal (1994) divided the recorded time histories of ground shaking during the Superstition Hills earthquake into four stages: Stage 1 (0.0–13.7 s): Ground acceleration was below ~0.1 g and pore water pressure buildup was small. Stage 2 (13.7–20.6 s): Strongest shaking occurred, with peak accelerations of 0.21 and 0.17 g at the surface and downhole instruments, respectively. Pore-water pressure increased rapidly, with small instantaneous drops. Stage 3 (20.6–40.0 s): Accelerations declined and stayed below 0.06 g. Pore-water pressure continued to increase at a high rate. Stage 4 (40.0–96.0 s): Ground acceleration was very low (~0.01 g), but excess pore pressure continued to rise, though at a slower rate, reaching the maximum pore pressure at 96 s. Thus a large portion of excess pore pressure at the Wildlife Reserve Array developed after the stronger, high-frequency ground motion had abated, and liquefaction did not occur until the earthquake was almost over (Holzer et al., 1989).

Many other investigations of the Wildlife recordings have been conducted (e.g., Zeghal and Elgamal, 1994; Youd and Carter, 2005; Holzer and Youd, 2007). Zeghal

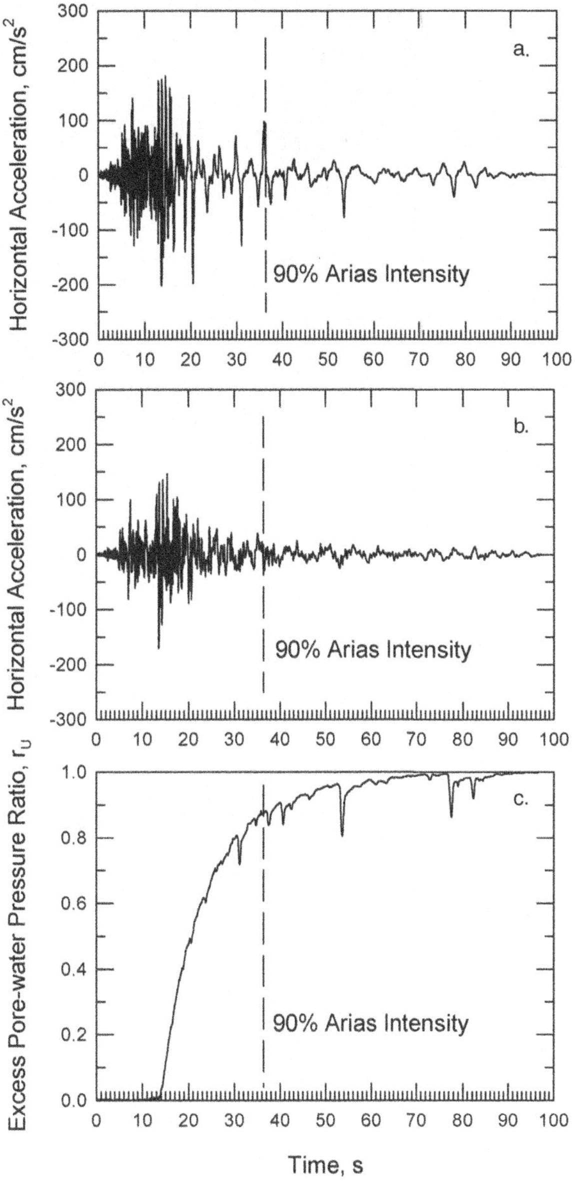

Fig. 2.9 Time histories of (**a**) north-south surface accelerations, (**b**) north-south downhole accelerations, and (**c**) excess pore pressure ratio recorded by piezometer P5 during and following the Superstition Hills earthquake. The downward spikes show rapid and transit decreases in pore-pressure. Ratio was calculated by dividing recorded values by the value at 97 s (From Holzer and Youd, 2007)

and Elgamal (1994) first demonstrated that the buildup of pore pressure was accompanied by a progressive softening of the sediments. Double-integrating the surface and downhole acceleration records leads to the time histories of displacements at the surface and at the downhole depth. The acceleration and displacement records may then be used to calculate the time histories of shear stress and the average shear strain (Zeghal and Elgamal, 1994). An example, recalculated by Holzer and Youd

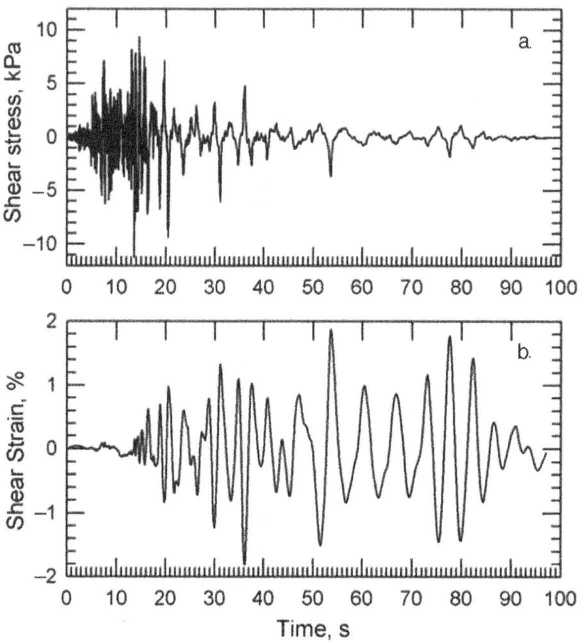

Fig. 2.10 Time histories of (**a**) north-south shear stress and (**b**) north-south shear strain at the Wildlife Reserve array during the Superstition Hills earthquake (From Holzer and Youd, 2007)

(2007), is shown in Fig. 2.10. An interesting result is that large amplitude (up to ~2%) long period (~5.5 s) cyclic shear strains continued to affect the sediments long after the high-frequency acceleration had abated. It shows that the sediments had softened so much that they underwent large shear deformations at very small shear stresses.

The progressive softening of sediments is best demonstrated by plotting the time history of shear stress against that of shear strain (Fig. 2.11), recalling that the slope of the stress-strain curve may be identified as the 'rigidity' of the sediments. At the onset of rapid pore-pressure increase, i.e., at 13.6 s (Fig. 2.9), the stress-strain curve shows steep slopes, i.e., high rigidity. With progressive increase in time, the slopes of the stress-strain curves rapidly decreased, showing that the sediments softened. Near the strain extremities, however, the slopes increase suddenly, showing that the sediments stiffened once more. This latter stiffening was attributed to strain-hardening (Zeghal and Engamal, 1994) and may be related to the rapid and transient decreases in pore pressure as recorded by the piezometers (Fig. 2.9; some of the decreases are labeled in Fig. 2.11), which, in turn, may be interpreted as a consequence of dilatancy in the strain-hardened sediments. With progressive softening, the activation of strain-hardening requires progressively greater amount of shear strain. As a result, large deformation may be induced by very small disturbances and the sediments fluidize.

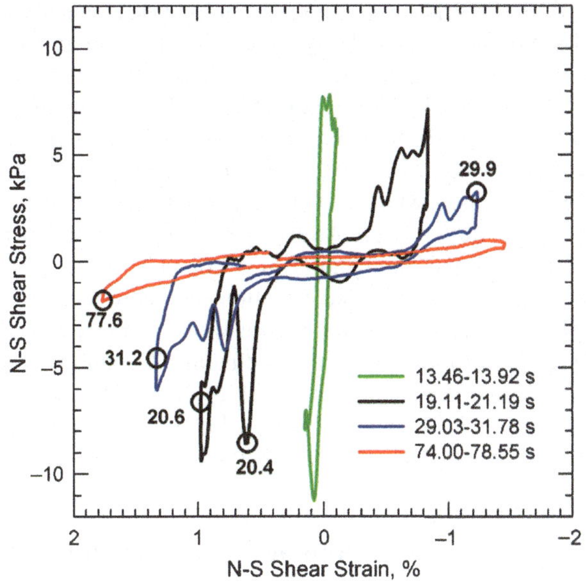

Fig. 2.11 Four hysteresis curves between shear stress and shear strain at different time segments. The times of instantaneous drop of pore pressure as recorded by piezometer P5 are labeled on the hysteresis curves (From Holzer and Youd, 2007)

It has been a challenge to explain why pore pressure continued to increase long after the ground acceleration had abated (e.g., Holzer et al., 1989; Holzer and Youd, 2007). One explanation is offered by the discussion in the last section. We note that the distance between the Wildlife Reserve Array and the epicenter of the M6.6 Superstition Hills earthquake (31 km, Holzer et al., 1989) is beyond the near field of the earthquake (<20 km); thus the seismic energy density at the Wildlife Reserve Array at the time of the earthquake may be too small to induce undrained consolidation, even in the most sensitive sediments. Second we note that the rise in pore pressure (Fig. 2.9c) was gradual and sustained, distinct from that caused by undrained consolidation which would have appeared as a steplike increase coincident with the strongest ground shaking (Roeloffs, 1998; Wang and Chia, 2008). The gradual and sustained change of pore pressure, however, can be readily explained by the diffusion of pore pressure from a nearby source that connected to the Wildlife Reserve Array through an earthquake- enhanced permeability, as discussed in the previous section. Under such condition, the duration of the pore-pressure increase does not depend upon the duration of ground shaking, but is rather a function of the distance between the pore-pressure source and the Wildlife Reserve Array as well as the permeability of the earth media between the two locations, thus explaining the continued pore-pressure buildup long after the ground acceleration had diminished. A different explanation offered by Holzer and Youd (2007) is that the strong ground shaking had initiated consolidation and thus pore-pressure increase in the sediments, and consolidation may have continued afterwards under the action of the long-period surface waves that arrived after the ground shaking had abated. If so, the sediments at the Wildlife Reserve Array would have to be more sensitive than the most sensitive sediments so far tested in the laboratory. An interesting point of this model is the positive feedback between pore-pressure buildup and sediment weakening, i.e., sediments which have been progressively weakened by rising pore pressure during seismic loading may continue to consolidate and generate pore pressure at progressively lower stresses, which further weakens the sediments.

In summary, the Wildlife Reserve Array experiment demonstrated that the occurrence of liquefaction is the culmination of a complex sequence of interactions among ground shaking, sediment deformation and pore-pressure redistribution and/or buildup. An increase in pore pressure weakens the sediment framework; this leads to greater deformation of the sediments. Continued increase in pore pressure may occur due to enhanced permeability connecting the sediments to a nearby source, or possibly by continued consolidation. This process continues at low frequency and very small shear stresses until the sediments liquefy.

2.6 Dependence of Liquefaction on Seismic Frequency

The period of seismic waves recorded near some liquefaction sites ranges from less than a second to many tens of seconds. Thus it is important to investigate whether the initiation of liquefaction depends on the frequency of seismic waves and, if so, how does it depend on the seismic frequency.

Established engineering methods frequently use the peak ground acceleration (PGA) as an index to predict liquefaction risk (Seed and Idriss, 1971). This is because PGA is proportional to the maximum shear stress induced in the sediment (Terzaghi et al., 1996). On the other hand, Midorikawa and Wakamatsu (1988) calculated PGA and PGV at $\sim$130 liquefaction sites and found that the occurrence of liquefaction is better correlated with the calculated PGV than with PGA. This result implies that liquefaction may be more sensitive to the low frequency components of the ground motion. This is because the integration of the acceleration records to calculate velocity filters out higher frequencies, so PGV is more dominated by low frequencies than PGA. In the following we test these models by using the occurrence of liquefaction, groundwater-level changes, and strong-motion records from central Taiwan during the Chi-Chi earthquake (Wang et al., 2003; Wong and Wang, 2007).

2.6.1 Field Observation from Taiwan

The 1999 $M_W 7.6$ Chi-Chi earthquake (Fig. 2.12) caused widespread liquefaction on the Choshui Alluvial Fan and the surrounding area (Fig. 2.13). An extensive network of strong-motion seismographs and a similarly extensive network of hydrologic monitoring wells were installed on the fan (Fig. 2.12) which captured both the ground motion and the concurrent groundwater level changes during and after the

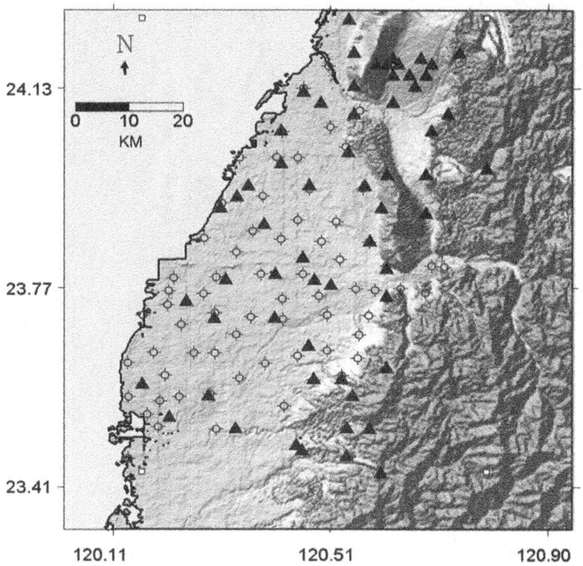

Fig. 2.12 Distribution of strong-motion stations (*solid triangles*) and hydrologic stations (*open circles*) on the Choshui alluvial fan (i.e., the flat fan-shaped area to the west of the hilly area) and nearby areas in western Taiwan. At each of the stations, there are one to five monitoring wells drilled to different depths up to 300 m. *Red star* marks the epicenter of the Chi-Chi earthquake, and *red curve* shows the ruptured fault in the earthquake (From Wang et al., 2006)

earthquake. These data provide a rare opportunity to investigate the field relationship among liquefaction, ground motion and groundwater level changes.

Taiwan is a north-south elongated island arc formed by the oblique collision between the Luzon volcanic arc on the Philippine Sea plate and the continental margin of China beginning in the late Cenozoic (Teng, 1990). The Choshui River Alluvial Fan (Fig. 2.12) is part of the Coastal Plain that lies along the western coast of the island and is covered by unconsolidated sediments of Neogene and Quaternary age, floored by a faulted basement. The Western Foothills that lie immediately to the east of the Coastal Plain, on the other hand, is a fold-and-thrust belt of consolidated sedimentary rocks, with virtually no loose sediments (Ho, 1988). The 1999 Chi-Chi ($M_w = 7.5$) earthquake, the largest to hit Taiwan in the last century, ruptured the Western Foothills along a ~80 km fault on the east of the Choshui River fan (Fig. 2.12).

Liquefaction was widespread in and near Yuanlin on the Choshui Alluvial Fan and further east along the ruptured fault (Fig. 2.13). The figure shows that the liquefaction sites on the Choshui River fan are closely associated with the largest coseismic rise of the groundwater level in the uppermost aquifer. No monitoring wells were installed in the basins east of the Choshui River fan; thus a similar comparison between pore pressure rise and the distribution of liquefaction cannot be made there.

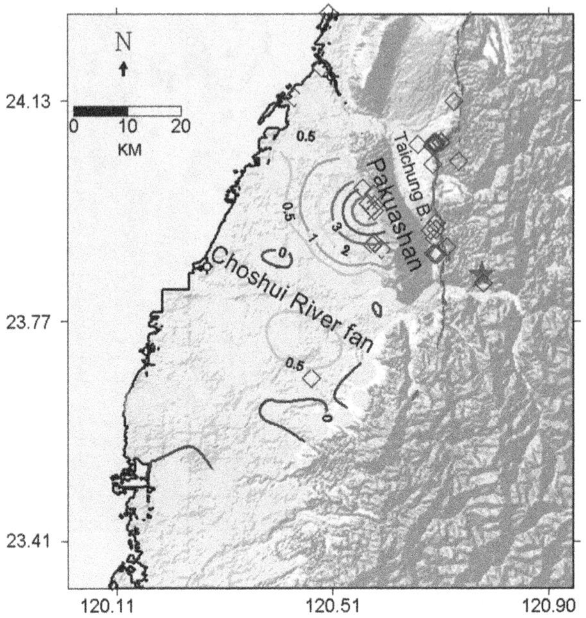

Fig. 2.13 Contours (in m) of the coseismic changes in groundwater level in the *topmost* aquifer in the Choshui alluvial fan. *Open diamond* show sites of liquefaction. Note that, on the Choshui alluvial fan, most liquefaction sites occurred in an area where the rise in groundwater level was above 2 m (Modified from Wang et al., 2006)

In order to test the frequency-dependence of pore-pressure development and liquefaction Wang et al. (2003) and Wong and Wang (2007) calculated the spectral acceleration, Sa, and spectral velocity, Sv, defined as the maximum response of a harmonic oscillator with a given damping coefficient with resonant frequency to the ground motion (Jennings, 1983). Sa and Sv were calculated at 5% damping at the location of each seismometer. Values for Sa and Sv throughout the region were then interpolated from the Sa and Sv values at the seismic stations (Fig. 2.13) using a kriging procedure. As an example, Fig. 2.14 shows maps of the spatial distribution of Sa at different frequencies together with the spatial distribution of the liquefaction sites. Visual inspection of the maps shows that there is a strong correlation between the liquefaction sites and Sa occurs at 0.5 and 1 Hz, but not at 2 Hz. A similar result occurs between the spatial distribution of Sv (not shown) and the liquefaction sites.

A more quantitative test of the above correlation of liquefaction with seismic wave frequency may be provided by plotting the t-values for the correlations of water level increase (i.e., pore pressure increase) with Sa and Sv over a range of frequencies in which liquefaction is mostly likely to occur. Calculations were made from $\sim 10^{-3}$ to $\sim 10^{2}$ Hz, but only a section of this range is shown in Fig. 2.15 for clarity. In general, Sa and Sv below about 0.8 Hz are more strongly correlated with the water-level increase than those above 0.8 Hz. The strength of the correlation peaks at 0.3–0.4 Hz, but declines rapidly at lower frequencies (Wong and Wang, 2007).

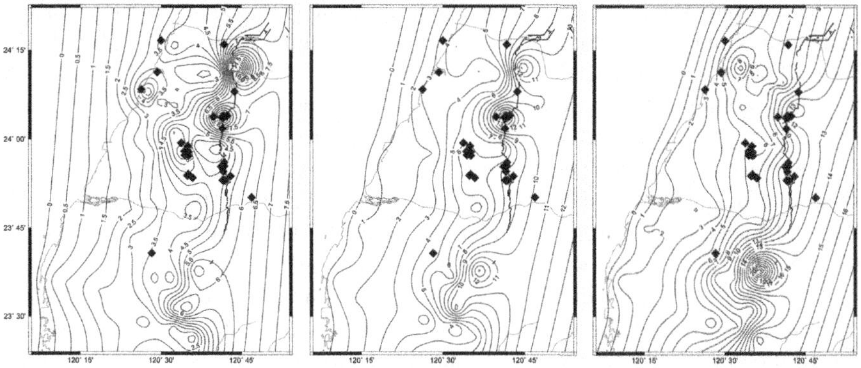

Fig. 2.14 Contours of Sa at 0.7 Hz (**a**), 1 Hz (**b**) and 2 Hz (**c**) during the Chi-Chi earthquake, plotted together with the distribution of liquefaction sites in solid diamonds. Note the strong correlation between liquefaction sites and Sa at 0.7 Hz and the weak correlation at 2 Hz (From Wong and Wang, 2007)

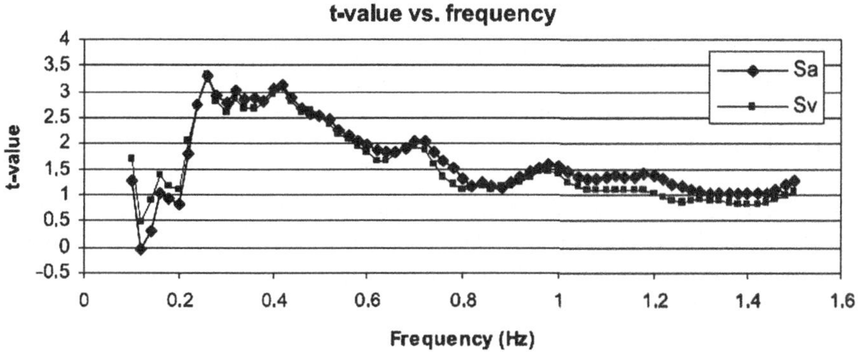

Fig. 2.15 t-values of the correlation of the water-level increase with Sa and Sv over a range of frequencies from 0.1 to 1.5 Hz, which pore pressure increases and liquefaction are typically attributed (From Wong and Wang, 2007)

2.6.2 Laboratory Studies

Only a few laboratory studies examined the dependence of liquefaction on the frequency of the seismic loads. Yoshimi and Oh-Oka (1975) conducted a series of cyclic shear tests under undrained conditions to determine the conditions to induce liquefaction in saturated sands. Most specimens in their experiments had a relative density, i.e., the ratio of the *density* of a specimen to the average *density* of the solid grains, of approximately 40%, and the frequency of the cyclic shear stress ranged from 1 to 12 Hz. They found that liquefaction failure became imminent when the ratio of the peak shear stress to the vertical effective stress reached a certain critical value, but the condition to induce liquefaction was nearly independent of the frequency of the cyclic shear stress from 1 to 12 Hz.

Sumita and Manga (2008) measured the rheology of non-Brownian particle suspensions under oscillatory shear at frequencies ranging from 0.1 to 10 Hz. A rheological transition was found to occur at a shear strain threshold of 10^{-4}, whereby the shear modulus of the viscoelastic suspension reduces sharply. This transition is in excellent correspondence with the threshold shear strain determined in geotechnical engineering experiments where excess pore pressure begins to develop and the shear modulus of the sediments begins to decline (Dobry et al., 1982; Vucetic, 1994; Hsu and Vucetic, 2004, 2006). Sumita and Manga (2008) found no dependence of the threshold shear strain on the frequency of shearing from 0.1 to 10 Hz.

Thus the field results and laboratory results on the dependence of liquefaction on frequency appear to be in conflict. On the one hand, existing laboratory results show little frequency-dependence of liquefaction; on the other hand, in situ studies of seismically instrumented liquefaction sites show an association of liquefaction with low-frequency ground motions.

2.6.3 Numerical Models

Using dynamic numerical models with nonlinear constitutive relations for sediments, Popescu (2002) and Ghosh and Madabhushi (2003) showed that the association of liquefaction and low-frequency ground motion may be due to sediment softening induced by ground motions. They also suggest a spectra-dependent feedback loop for liquefying sediments: low frequency excitation causes ground softening and pore pressure increases more efficiently than for high frequency excitation. This softening in turn reduces the resonant frequency of the sediment column, amplifying low frequency motions and damping high frequency motions, leading to further softening and pore pressure increases, possibly leading to liquefaction.

Kostadinov and Towhata (2002) proposed a linearly elastic model of one dimensional wave propagation that suggests liquefaction may occur when the sediment column reaches a resonant state. Similarly, Bachrach (2001) used a dynamic poroelastic model to simulate the effect of P-waves on pore-pressure buildup and liquefaction near the resonant frequency of sediment columns.

Further in situ, laboratory, and theoretical work are required to evaluate the dependence of pore-pressure buildup and liquefaction on the frequency of seismic waves. If the frequency dependence is due to resonance in the soil, as theoretical models suggest, local hydrologic and geologic conditions would affect ground motion frequencies.

The roles that different types of seismic waves play in inducing liquefaction also needs to be better investigated. Finally, to make predictions regarding liquefaction at particular sites, results must be integrated with site-specific geotechnical data. This requires the development of predictive theories of liquefaction that incorporate both the seismic spectral information of the ground motion, as well as geotechnical information such as SPT and CPT. Such predictions should be verified with data from earthquake-affected sites where both geotechnical data and ground motion data are available.

2.7 Concluding Remarks

Earthquake-induced liquefaction has been studied by earthquake engineers based on Terzaghi's concept that consolidation of loose sediments raises pore pressure which eventually causes liquefaction. Here we show that, while this mechanism may be valid in the near field of an earthquake, the energy of seismic waves at distances beyond the near field may be too small to induce consolidation, even in the most sensitive sediments. Hence a new mechanism may be needed to explain the abundant occurrences of liquefaction beyond the near field. Here we proposed a redistribution of pore pressure due to earthquake-enhanced permeability as a mechanism to explain these occurrences. The proposed mechanism is supported by evidence that pore pressure in the field is often heterogeneous at a local scale and that seismic waves can enhance the permeability of shallow crust at distances far beyond the near field. Thus an enhanced permeability during an earthquake may connect sites of different pore pressures in the shallow crust, which were hydraulically isolated from each other before the earthquake, allowing pore pressure to redistribute. This redistribution may raise the pore pressure at some sites to facilitate liquefaction.

An unresolved issue is the complex relationship between liquefaction and the frequency of seismic waves. Current results from the field and laboratories are in conflict. Future work is needed to resolve these conflicts.

References

Ambraseys, N.N., 1988, Engineering seismology, *Earthquake Eng. Structural Dynamics, 17*, 1–105.

Bachrach, R., A. Nur, and A. Agnon, 2001, Liquefaction and dynamic poroelasticity in soft sediments, *J. Geophys. Res., 106*, 13515–13526.

Bath, M., 1966, Earthquake energy and magnitude, *Phys. Chem. Earth, 7*, 115–165.

Berrill, J.B., and R.O. Davis, 1985, Energy dissipation and seismic liquefaction in sands: Revised model, *Soils Found., 25*, 106–118.

Brodsky, E.E., E. Roeloffs, D. Woodcock, I. Gall, and M. Manga, 2003, A mechanism for sustained water pressure changes induced by distant earthquakes, *J. Geophys. Res., 108*, doi:10.1029/2002JB002321.

Cua, G.B., 2004, *Creating the Virtual Seismologist: Developments in Ground Motion Characterization and Seismic Early Warning*, Ph.D. Dissertation, Caltech.

Dief, H.M., 2000, *Evaluating the Liquefaction Potential of Soils by the Energy Method in the Centrifuge*, Ph.D. Dissertation, Department of Civil Engineering, Case Western Reserve University, Cleveland, OH.

Dobry, R., R.S. Ladd, F.Y. Yokel, R.M. Chung, and D. Powell, 1982, Prediction of pore water pressure buildup and liquefaction of sands during earthquakes by the cyclic strain method, *National Bureau of Standards Building Science Series, 138*, National Bureau of Standards and Technology, Gaithersburg, MD, pp. 150.

Elkhoury, J.E., E.E. Brodsky, and D.C. Agnew, 2006, Seismic waves increase permeability, *Nature, 411*, 1135–1138.

Figueroa, J.L., A.S. Saada, L. Liang, N.M. Dahisaria, 1994, Evaluation of soil liquefaction by energy principles, *J. Geotech. Eng., ASCE, 120*, 1554–1569.

Galli, P., 2000, New empirical relationships between magnitude and distance for liquefaction, *Tectonophysics, 324*, 169–187.

Ghosh, B. and S.P.G. Madabhushi, 2003, A numerical investigation into effects of single and multiple frequency earthquake motions, *Soil Dynamics and Earthquake Engineering, 23*, 691–704.

Green, R.A., and J.K. Mitchell, 2004, Energy-based evaluation and remediation of liquefiable soils. In: M. Yegian, and E. Kavazanjian (eds.), *Geotechnical Engineering for Transportation Projects, ASCE Geotechnical Special Publication, No. 126, Vol. 2*, 1961–1970.

Hazirbaba, K., and E.M. Rathje, 2004, A comparison between in situ and laboratory measurements of pore water pressure generation. In: *13th World Conference on Earthquake Engineering, paper no. 1220*, Vancouver.

Ho, C.S., 1988, *An Introduction to the Geology of Taiwan*, 2nd ed., 192 pp, Taiwan: Central Geological Survey.

Holzer, T.L., J.C. Tinsley, and T.C. Hank, 1989, Dynamics of liquefaction during the 1987 Superstition Hills, California, earthquake, *Science, 244*, 56–59.

Holzer, T.L. and T.L. Youd, 2007, Liquefaction, ground oscillation, and soil deformation at the Wildlife Array, California, *Bull. Seis. Soc. Am., 97*, 961–976.

Hsu, C.C., and M. Vucetic, 2004, Volumetric threshold shear strain for cyclic settlement, *J. Geotech. Geoenviron. Eng., 130*, 58–70.

Hsu, C.C., and M. Vucetic, 2006, Threshold shear strain for cyclic pore-water pressure in cohesive soils, *J. Geotech. Geoenviron. Eng., 132*, 1325–1335.

Jennings, P.C., 1983, Engineering seismology. In: H. Kanamori, and E. Boschi (ed.), *Earthquakes: Observation, Theory and Interpretation*, Amsterdam: North-Holland.

Kokusho, T., 2007, Liquefaction strengths of poorly-graded and well-graded granular soils investigated by lab tests. In: K.D. Pitilakis (ed.), *Earthquake Geotechnical Engineering*, Dordrecht: Springer.

Kostadinov, M.V., and I. Towhata, 2002, Assessment of liquefaction-inducing peak ground velocity and frequency of horizontal ground shaking at onset of phenomenon, *Soil Dyn. Earthq. Eng., 22*, 309–322.

Kuribayashi, E., and F. Tatsuoka, 1975, Brief review of liquefaction during earthquakes in Japan, *Soils Found., 15*, 81–92.

Law, K.T., Y.L. Cao, and G.N. He, 1990, An energy approach for assessing seismic liquefaction potential, *Can. Geotech. J., 27*, 320–329.

Lay, T., and T.C. Wallace, 1995, *Modern Global Seismology*, pp. 521, San Diego: Academic Press.

Liang, L., J.L. Figueroa, and A.S. Saada, 1995, Liquefaction under random loading: Unit energy approach, *J. Geotech. Eng., 121*, 776–781.

Marinatos, S.N., 1960, Helice: A submerged town of classical Greece, *Archaeology, 13*, 186–193.

Midorikawa, S., and K. Wakamatsu, 1988, Intensity of earthquake ground motion at liquefied sites, *Soils Found., 28*, 73–84.

National Research Council, 1985, *Liquefaction of Soils during Earthquakes*, pp. 240, National Academy Press, Washington, DC.

Nemat-Nasser, S., and A. Shokooh, 1979, A unified approach for densification and liquefaction of cohesionless sands in cyclic loading, *Can. Geotech. J., 16*, 659–678.

Obermeir, S.F., 1989, The new madrid earthquakes: An engineering-geologic interpretation of relic liquefaction features, *U.S. Geol. Surv. Prof. Pap., 1336-B*, 114.

Papadopoulos, G.A., and G. Lefkopulos, 1993, Magnitude-distance relations for liquefaction in soil from earthquakes, *Bull. Seism. Soc. Am., 83*, 925–938.

Pitilakis, K.D. (ed.), 2007, *Earthquake Geotechnical Engineering*, Dordrecht: Springer.

Popescu, R., 2002, Finite element assessment of the effects of seismic loading rate on soil liquefaction, *Canadian Geotech. J., 29*, 331–334.

Roeloffs, E.A., 1998, Persistent water level changes in a well near Parkfield, California, due to localand distant earthquakes, *J. Geophys. Res., 103*, 869–889.

Schmidt, F.J., 1875, *Studien uber Erdbeben*, pp. 360, Leipzig: Carl Scholtze.

Seed, H.B., 1968, Landslides during earthquakes due to soil liquefaction, *J. Soil Mech. Found. Div., 94*, 1055–1122.

Seed, H.B., and K.L. Lee, 1966, Liquefaction of saturated sands during cyclic loading, *J. Soil Mech. Found. Div., 92*, 105–134.

Seed, H.B., and I.M. Idriss, 1967, Analysis of soil liquefaction: Niigata earthquake, *J. Soil Mech. Found. Div., 93*, 83–108.

Seed, H.B., and I.M. Idriss, 1971, Simplified procedure for evaluating soil liquefaction potential, *J. Soil Mech. Found. Div, 97*, 1249–1273.

Su, T.-C., K.-W. Chiang, S.-J. Lin, F.-G. Wang, and S.-W. Duann, 2000, Field reconnaissance and preliminary assessment of liquefaction in Yuan-Lin area, *Sino-Geotechnics, 77*, 29–38.

Sumita, I., and M. Manga, 2008, Suspension rheology under oscillatory shear and its geophysical implications, *Earth Planet. Sc. Lett., 269*, 467–476.

Teng, L.S., 1990, Geotectonic evolution of late Cenozoic arc-continent collision in Taiwan, *Tectonophysics, 183*, 57–76.

Terzaghi, K., 1925, *Erdbaummechanic,* Vienna: Franz Deuticke.

Terzaghi, K., R.B. Peck, and G. Mesri, 1996, *Soil Mechanics in Engineering Practice*, 3rd ed., pp. 195, New York: John Wiley and Sons.

Tuttle, M.P., and E.S. Schweig, 1996, Recognizing and dating prehistoric liquefaction features: Lessons learned in the New Madrid seismic zone, central United States, *J. Geophys. Res., 101*, 6171–6178.

Vucetic, M., 1994, Cyclic threshold of shear strains in soils, *J. Geotech. Eng., 120*, 2208–2228.

Waller, R.M., 1966, Effects of the March 1964 Alaska earthquake on the hydrology of south-central Alaska, *U.S. Geol. Surv. Prof. Pap., 544-A*.

Wang, C.-Y., 2007, Liquefaction beyond the near field, *Seismo. Res. Lett., 78*, 512–517.

Wang, C.-Y., and Y. Chia, 2008, Mechanism of water level changes during earthquakes: Near field versus intermediate field, *Geophys. Res. Lett., 35*, L12402, doi:10.1029/2008GL034227.

Wang, C.-Y., D.S. Dreger, C.-H. Wang, D. Mayeri, and J.G. Berryman, 2003, Field relations among coseismic ground motion, water level change, and liquefaction for the 1999 Chi-Chi ($M_w = 7.5$) earthquake, Taiwan, *Geophys. Res. Lett., 30*, 1890, doi:10.1029/2003GL017601.

Wang, C.-Y., A. Wong, D.S. Dreger, and M. Manga, 2006, Liquefaction limit during earthquakes and underground explosions – implications on ground-motion attenuation, *Bull. Seis. Soc. Am., 96*, 355–363.

Water Resource Bureau, 1999, *Summary Report of Groundwater Monitoring Network Plan in Taiwan, Phase I, Hydrogeology of Choshui River Alluvial Fan*, Water Resource Bureau, Ministry of Economic Affairs, Taipei, Taiwan, 240 pp (in Chinese).

Wells, D.L., and K.J. Coppersmith, 1994, New empirical relationships among magnitude, rupture length, rupture width, rupture area, and surface displacement, *Bull. Seis. Soc. Am., 84*, 974–1002.

Wong, A., and C.-Y. Wang, 2007, Field relations between the spectral composition of ground motion and hydrological effects during the 1999 Chi-Chi (Taiwan) earthquake. *J. Geophys. Res., 112*, B10305, doi:10.1029/2006JB004516.

Yoshimi, Y., and H. Oh-Oka, 1975, Influence of degree of shear stress reversal on the liquefaction potential of saturated sand, *Soils Found. (Japan), 15*, 27–40.

Youd, T.L., 1972, Compaction of sands by repeated shear straining, *J. Soil Mech. Found. Div., Am. Soc. Civ. Eng., 98*, 709–725.

Youd, T.L., E.L. Harp, D.K. Keefer, and R.C. Wilson, 1985, The Borah Peak, Idaho, earthquake of October 28, 1983 – Liquefaction, *Earthquake Spectra, 2*, 71–89.

Youd, T.L., and B.L. Carter, 2005, Influence of soil softening and liquefaction on spectral acceleration, *J. Geotech. Geoenviron. Eng., 131*, 811–825.

Zeghal, M., and A.-W. Elgamal, 1994, Analysis of site liquefaction using earthquake records, *J. Geotech. Eng., 120*, 996–1017.

Chapter 3
Mud Volcanoes

Contents

3.1 Introduction

Mud volcanoes are surface structures formed by the eruption of mud from the subsurface. Figure 3.1 shows a typical example. The erupted materials are usually fine grained sediment, water, and gases, dominantly CO_2 and methane. Fragments of country rock are also sometimes entrained. They range in size from <1 m, typical of mud volcanoes formed by liquefaction at shallow depths, to edifices that are a few hundred meters high or extend laterally for more than 1 km. The largest eruptions deliver mud to the surface from depths that exceed more than 1 km (Kopf, 2002). In this chapter we focus on the larger eruptions; small eruptions that originate in the shallowest subsurface are manifestations of the 'traditional' liquefaction discussed in Chap. 2.

Mud volcanism requires thick layers of unconsolidated sediment with high pore pressures. They are thus most common in areas with high sedimentation rates such as sedimentary basins and accretionary prisms. Their eruption is favored by compressional settings which act to increase pore pressures.

C.-Y. Wang, M. Manga, *Earthquakes and Water*, Lecture Notes in Earth Sciences 114, DOI 10.1007/978-3-642-00810-8_3, © Springer-Verlag Berlin Heidelberg 2010

Fig. 3.1 Garadag mud volcano, 40 km south of Baku, Azerbaijan . Photo taken by Martin Hovland, and reproduced with permission

3.2 Response of Mud Volcanoes to Earthquakes

The number of documented triggered mud volcano eruptions is small. Mellors et al. (2007) compiled global reports and observations and identified 13 triggered eruptions within days of earthquakes. Manga et al. (2009) identify two more events. Of the grand total, 6 instances are triggered eruptions at the Niikappu mud volcano in Japan (Chigira and Tanaka, 1997). Another 5 are from Azerbaijan (Aliyev et al., 2002).

Figure 3.2 shows the relationship between earthquake magnitude and the distance between triggered eruptions of mud volcanoes and the earthquake epicenter. Also shown for reference is the empirical distance threshold from Chap. 2 for the occurrence of liquefaction. Although the number of triggered events is small, mud volcanoes appear to exhibit a sensitivity to earthquakes consistent with that for much more shallow liquefaction.

The repeated eruption of the Niikappu mud volcanoes, Japan, in response to earthquakes offers an excellent opportunity to better understand the conditions required for triggering. This is analogous to studying the response of a single water well to multiple earthquakes (Chap. 5). Manga et al. (2009) found that this mud volcano consistently obeyed the empirical threshold in Fig. 3.2 provided there was a repose time of at least 1–2 years between eruptions. Large, close earthquakes that occurred sooner did not trigger an eruption. This supports the arguments in Mellors et al. (2007) that a recharge period is needed before another eruption can be triggered.

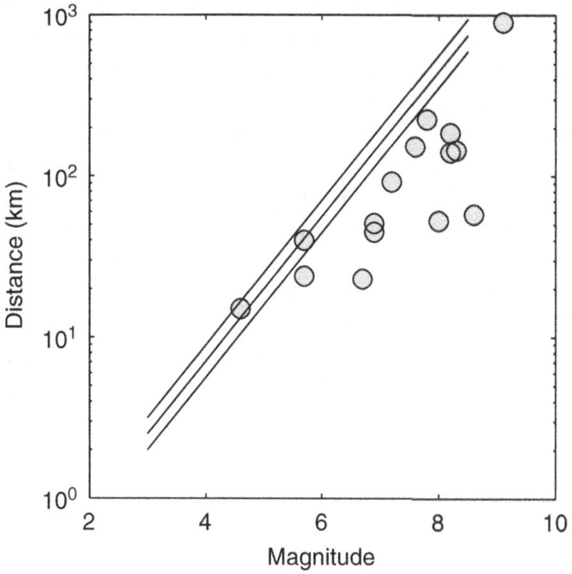

Fig. 3.2 Relationship between earthquake magnitude and distance of mud volcanoes that erupted within days of the earthquake. Data from Mellors et al. (2007), Manga and Brodsky (2006), and Manga et al. (2009). The sloping lines show the mean (and error) of the maximum distance for liquefaction observations (Wang, 2007)

3.3 Insights from Triggered Eruptions of Magmatic Volcanoes

The small number of well-documented eruptions of mud volcanoes limits our ability to perform a meaningful analysis of the probability that they are triggered by earthquakes. This is less problematic for earthquake-triggered eruptions of magmatic ('real') volcanoes. We thus provide a brief overview of what is known about the triggered eruptions of volcanoes and implications for mud volcanoes. Reviews of this topic are published by Hill et al. (2002) and Manga and Brodsky (2006).

The Smithsonian Institution maintains a catalog of volcanic eruptions that includes the date and magnitude of volcanic eruptions (Siebert and Simkin, 2002; www.volcano.si.edu/world). This catalog is statistically complete and meaningful for moderate to large explosive eruptions since about 1500 AD. It is thus possible to look for correlations between the occurrence of large (magnitude >8) earthquakes and volcanic eruptions over a period of at least 500 years, and with smaller earthquakes during the more recent past. In regions with a longer recorded history, e.g., Italy, a regional analysis may permit analysis of smaller earthquakes and eruptions extending further back in time.

Identifying a triggered eruption suffers from the complication that the surface manifestation of a triggered event may not occur for days to perhaps even years after the earthquake. The nature of any delay reflects the mechanism of triggering and the manner in which the magma erupts. The search for triggered eruptions is thus generally confined to a specific window in space and time. For example, Linde and Sacks (1998) showed that more eruptions occurred within a couple days of large earthquakes than can be expected by chance. Their analysis was updated by Manga and Brodsky (2006), and is shown in Fig. 3.3 for eruptions with magnitude VEI $\geq$ 2 and within 5 fault lengths of the earthquake. VEI is the Volcanic Explosive Index, and a value of 2 corresponds to moderate explosive eruptions (Newhall and Self, 1982). If the 10 events that occur within days of the earthquakes is an accurate assessment of the number of triggered eruptions, this represents 0.4% of eruptions of magnitude VEI $\geq$ 2. Lemarchand and Grasso (2007) performed a similar analysis that included both smaller earthquakes and eruptions for the period 1973–2005 and similarly found that 0.3% of eruptions interacted with earthquakes (though for these smaller events, the occurrence of eruptions is distributed approximately symmetrically in time around the earthquake).

The aforementioned studies focused on eruptions within days for which a statistical analysis is easier to perform (Linde and Sacks, 1998). Delayed triggering is

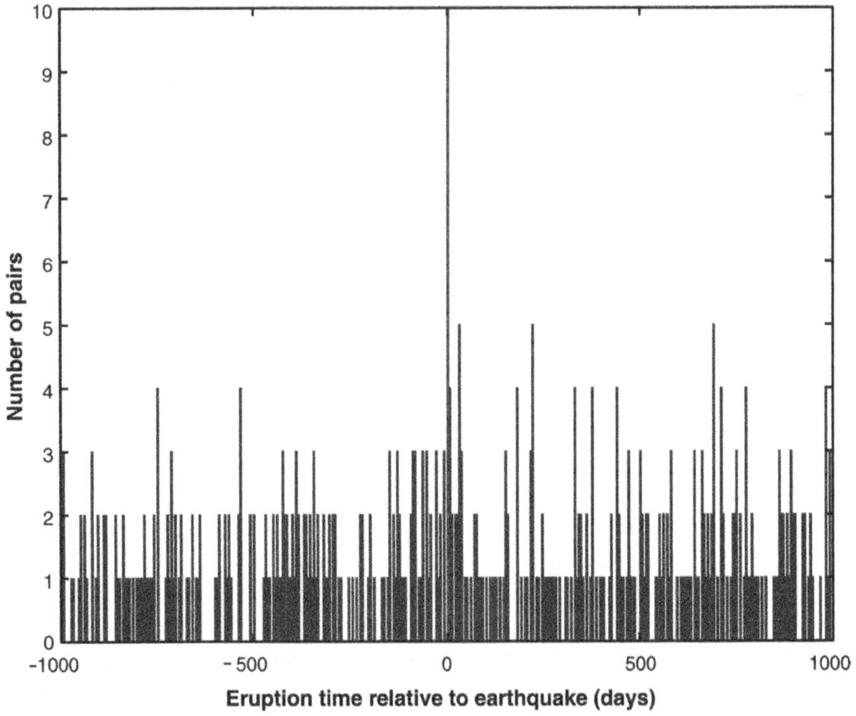

Fig. 3.3 Histogram showing the relationship between the time of volcanic eruptions relative to the time of large (magnitude > 8) earthquakes. Bin size is 5 days. The peak corresponds to eruptions within 5 days of the earthquakes, and most of these are (statistically likely) to be triggered eruptions (Figure from Manga and Brodsky, 2006)

more difficult to establish and several studies have examined the space-time connections between earthquakes and eruptions (e.g., Marzocchi, 2002; Watt et al., 2009). Proposed examples of delayed triggering include an increase in eruptions in the Cascade arc, USA, in the 1800s following a large subduction earthquake in 1700 (Hill et al., 2002); volcanic eruptions following the M9.3 December 2004 Sumatra earthquake, the 1952 M9.0 Kamchatka earthquake, and the 1964 M9.2 Alaska earthquake (Walter and Amelung, 2006); increased eruption rates after Chilean earthquakes, 1906 M8 and 1964 M9.5 (Watt et al., 2009); the 1991 eruption of Pinatubo 11 months after a M7.7 event (Bautista et al., 1996).

3.4 Mechanisms

The mechanisms that trigger magmatic eruption are likely to be more difficult to identify than the mechanisms that account for hydrological responses. This is because there are a greater number and complexity of processes that operate within magma chambers and influence the ascent of magma. Here we review some of the mechanisms that have been proposed as triggers for both mud and magmatic volcanoes.

3.4.1 Static or Dynamic Stresses?

A central theme in studies of triggered eruptions is whether the triggering is controlled by static or dynamic stress changes. Manga and Brodsky (2006) argue that the static stress changes caused by earthquakes are in general too small to initiate eruption through any mechanism, and favor processes that are able to turn larger amplitude dynamic strains into some type of permanent or semi-permanent change.

In support of an important role of static stress changes, Walter and Amelung (2006) document a systematic pattern of coseismic volumetric expansion at triggered volcanoes. It is not intuitive that the pressure decrease in magma chambers that would accompany volumetric expansion would promote eruption: eruption should require an overpressure to force magma out of the chamber, or to create new dikes.

3.4.2 Mechanisms for Initiating Eruptions

Magmatic and mud volcanoes share in common that gases play a role in providing buoyancy, they erupt materials that are liquefied or fluidized, and the source is usually overpressured. Mechanisms through which dynamic strains influence the nucleation or growth of bubbles, or liquefy sediment or crystal mushes, are in principle possible in both systems.

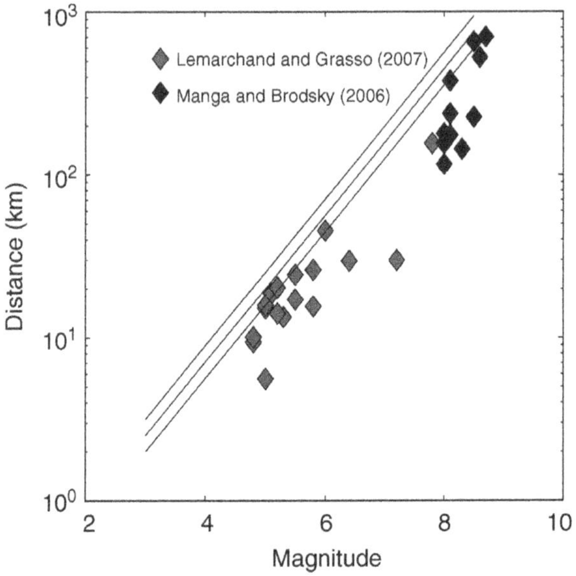

Fig. 3.4 Relationship between earthquake magnitude and distance of triggered volcanic eruptions. Data come from compilations in Manga and Brodsky (2006) for magnitude > 8 earthquakes, and Lemarchand and Grasso (2007) for smaller earthquakes. Only eruptions within 5 days of the earthquake are considered. The sloping lines indicate the maximum distance over which liquefaction has been documented (Wang, 2007)

Figure 3.4 summarizes the relationship between earthquake magnitude and the distance from the epicenter of triggered eruptions of magmatic volcanoes. Only eruptions triggered within days are included (data from Manga and Brodsky, 2006; Lemarchand and Grasso, 2007). For reference, the threshold for other hydrological responses discussed in previous chapters is also shown. Remarkably, the threshold distance is similar for all these phenomena, magmatic and mud volcanoes included.

3.4.2.1 Mechanisms Involving Bubbles

Given the importance of bubbles in driving magma to the surface and powering volcanic eruptions, several triggering mechanisms have been proposed that invoke bubbles. One possibility is the nucleation of new bubbles in a supersaturate liquid by the periodic changes in pressure generated by seismic waves (e.g., Manga and Brodsky, 2006). A second possibility, is that diffusion of gas from a supersaturated liquid into preexisting bubbles is enhanced by dynamic strains. When bubble experience oscillatory strain, there is an asymmetric diffusion of gas into and away from the bubble owing to the change in shape – this process is called rectified diffusion (e.g., Sturtevant et al., 1996). Recent work by Ichihara and Brodsky (2006) has shown that this process results in insignificant growth of bubbles. A third possibility, is that pore pressures rise as bubbles carry high pressures to shallower depths as they rise (e.g., Sahagian and Proussevitch, 1992; Linde et al., 1994), a process called advective overpressure. This mechanism requires that both the bubbles and surrounding matrix/liquid are incompressible, and several studies have shown that these assumptions are not satisfied (Bagdassarov, 1994; Pyle and Pyle, 1995). A fourth possibility is that gas hydrates dissociate. Submarine mud volcanoes are often associated with gas hydrates (Milkov, 2000) and enhanced methane emission has been attributed to earthquakes in both lakes (Rensbergen et al., 2002) and the ocean (Mau et al., 2007). However, triggered eruptions that have been identified so far are subaerial (this is very likely an observational bias), where gas hydrates should not exist.

One intriguing possibility is that the volume expansion of the magma chamber can lead to a net increase in its overpressure owing to the growth of bubbles. Recall that Walter and Amelung (2006) found that triggered magmatic eruption occur in regions that experience volumetric expansion. Nishimura (2004) showed that the growth of bubbles that accompanies magma chamber expansion, causes a decrease in pressure difference between that inside and outside bubbles, and the surface tension energy liberated results in a net pressure increase in the magma. This effect is very small, except for very small bubbles (smaller than a few microns).

3.4.2.2 Liquefaction

As mud volcanoes erupt liquefied or fluidized sediment, mechanisms that invoke liquefaction by dynamic strain seem reasonable. However, liquefaction is generally viewed as a shallow phenomenon because overburden stresses at greater depths require that pore pressure changes become unrealistically large (e.g., Youd et al., 2004; Chap. 2). This should not be a limitation in the settings where mud volcanism occurs as the erupted materials initially had high pore pressures, and only modest increases in pore pressure may be necessary even if the overburden stresses are high.

Liquefaction or weakening of magmatic suspensions has also been invoked to explain the seismic triggering of magmatic volcanoes (Hill et al., 2002; Sumita and Manga, 2008). There is no observational evidence that supports this suggestion.

3.4.2.3 Breaching Reservoirs

Water level changes in wells, as shown in Fig. 2.7 and discussed later in Chap. 5, can be explained in many instances by changes in permeability or the breaching of hydrological barriers that allow fluids and pore pressure to be redistributed. This is a viable mechanism to fluidize or liquefy unconsolidated sediments if there are nearly reservoirs with high enough pore pressure (e.g., Wang, 2007). In some settings, the gases that erupt at mud volcanoes may be sourced much deeper than the erupted mud (e.g., Cooper, 2001), supporting the idea that fluid and gas migration play a role in initiating eruptions. This process may induce, however, a time lag in the manifestation of the triggered eruption governed by the time scale for fluids and/or gas to migrate (e.g., Husen and Kissling, 2001).

3.5 Effect of Earthquakes on Already-Erupting Mud Volcanoes

Given the strong sensitivity of geysers – also already-erupting systems – to earthquakes, we can reasonably expect already-erupting mud volcanoes to be more sensitive to earthquake than the triggering of new eruptions. The limited observations of magmatic volcanoes support this contention. Harris and Ripepe (2007) report changes in eruption rate at Semaru volcano, Indonesia in response to the 2006 M6.3 Yogyakarta earthquake based on satellite thermal imaging. The magnitude and distance of this event also place it above the empirical triggering threshold in Fig. 3.4.

Compared with geysers, there are fewer robust and quantitative data documenting how already erupting mud volcanoes respond to earthquakes. In part this reflects the difficult in quantifying eruption rate. Anecdotal accounts, however, abound.

The LUSI mud volcano in Indonesia, pictured in Fig. 3.5, that began erupting in May 2006 offers perhaps the best record of how an active mud volcano responds to earthquakes because it erupted for a long period of time in a very seismically active area. Because of its location in a densely populated region, its eruption was closely monitored. Mazzini et al. (2007) suggest that some of the flow increases are correlated with regional earthquakes. They noted that not all changes in eruption rate or style are correlated with earthquake. Davies et al. (2008) report anecdotal accounts of responses to larger and more distant earthquakes, and these responses occur for earthquakes too distant to be expected to trigger a new eruption (these events plot above the threshold shown in Fig. 3.2).

Manga et al. (2009) undertook a more systematic study of the response of LUSI. Figure 3.6 shows the date of regional earthquakes and a proxy for the magnitude of shaking (distance divided by the square root of fault area) along with the eruption

Fig. 3.5 Photo of the LUSI mud volcano, Indonesia. *Left*: Satellite image from May 5, 2008 almost two years after eruption (area shown is 3.7 by 4.0 km). *Right*: The mud is being contained by dikes and pumped into the Porong river (photo provided by Richard Davies). It has been proposed that this eruption was triggered by the 2006 M6.3 Yogyakarta earthquake 250 km away (Mazzini et al., 2007), though this trigger is disputed (Davies et al., 2007, 2008; Manga, 2007). Over the first two years of eruption, the eruption rate averaged more than 100,000 m^3/day. More than 30,000 people have been displaced

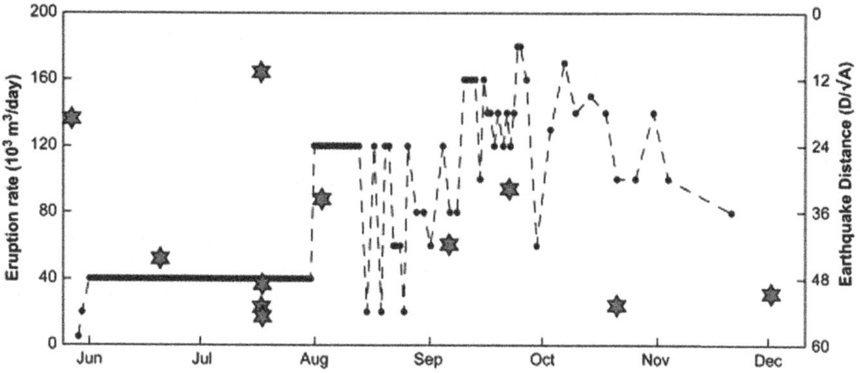

Fig. 3.6 Eruption rate at Lusi mud volcano. Data provided by A. Mazzini. *Dashed lines* connect actual measurements. The *red stars* scale with the expected magnitude of ground shaking – distance from the epicenter normalized by the square root of the rupture area (which scales with earthquake magnitude) (From Manga et al., 2009)

rate reported in Mazzini et al. (2007). Figure 3.6 shows no systematic relationship between the shaking caused by earthquakes and any changes in eruption rate, but the temporal sampling of eruption rate and accuracy of measurements providing limit the ability to draw reliable conclusions.

3.6 Concluding Remarks

It is important to recall that most eruptions at mud volcanoes (Mellors et al., 2007), and at magmatic volcanoes (Linde and Sacks, 1998; Manga and Brodksy, 2006), are not triggered by earthquakes. This implies that for triggered eruptions the plumbing system for the mud or magma must already be near failure, perhaps with stresses within less than 5% of the failure stress (Manga et al., 2009). If we choose this failure stress to be the tensile strength of rock, say 10 MPa, extra stresses of only 0.5 MPa are needed.

It is clear that the number of triggered events is too small, and the amount and quality of data from erupting mud volcanoes too limited, to conclusively answer the most interesting questions about triggered eruptions: static or dynamic stress triggering? mechanism? is there increased earthquake sensitivity once the eruption begins? Key to addressing questions about triggering are more examples, accurate timing, and ideally nearly co-located eruptions and seismometers. Both mud and volcanic eruptions begin in the subsurface, and seismic and deformation signal that accompany the initiation of unrest prior to the surface expression of eruption may be critical for identifying the mechanisms that lead to eruption. For already-erupting mud volcanoes, continuous gas flux measurements (e.g., Yang et al., 2006) or continuous GPS (e.g., Abidin et al., 2006) are promising approaches that offer high temporal resolution.

References

Abidin, H.Z., R.J. Davies, M.A. Kusuma, H. Andreas, and T. Deguchi, 2008, Subsidence and uplift of Sidoarjo (east Java) due to the eruption of the Lusi mud volcano (2006-present), *Environ. Geol.*, doi:10.1007/s00254-008-1363-4.

Aliyev, A.A., I.S. Guliyev, and I.S. Belov, 2002, *Catalogue of Recorded Eruptions of Mud Volcanoes of Azerbaijan for a Period of Years 1810–2001*, 88 pages, Baku, Azerbaijan: Nafta Press.

Bagdassarov, N., 1994, Pressure and volume changes in magmatic systems due to the vertical displacement of compressible materials, *J. Volcanol. Geotherm. Res., 63*, 95–100.

Bautista, B.C. et al., 1996, Relationship of regional and local structures to Mount Pinatubo activity. In: C. Newhall, and R.S. Punongbayan (ed.), *Fire and Mud*, Seattle: University of Washington Press.

Chigira, M., and K. Tanaka, 1997, Structural features and the history of mud volcanoes in Southern Hokkaido, Northern Japan, *J. Geol. Soc. (Japan), 103*, 781–791.

Cooper, C., 2001, Mud volcanoes of Azerbaijan visualized using 3D seismic depth cubes: The importance of overpressured fluid and gas instead of non-existed diapirs. In: *Conference on Subsurface Sediment Mobilization.* Ghent, Belgium: European Association of Geoscientists and Engineers.

Davies, R.J., R.E. Swarbrick, R.J. Evans, and M. Huuse, 2007, Birth of a mud volcano: East Java, 29 may 2006, *GSA Today, 17*, 4–9.

Davies, R.J., M. Brumm, M. Manga, R. Rubiandini, and R. Swarbrick, 2008, The east Java mud volcano (2006-present): An earthquake or drilling trigger? *Earth Planet. Sci. Lett., 272*, 627–638.

Harris, A.J.L., and M. Ripepe, 2007, Regional earthquake as a trigger for enhanced volcanic activity: Evidence from MODIS thermal data, *Geophys. Res. Lett., 34*, L02304.

Hill, D.P., F. Pollitz, and C. Newhall, 2002, Earthquake-volcano interactions. *Phys. Today, 55*, 41–47.

Husen, S., and E. Kissling, 2001, Postseismic fluid flow after the large subduction earthquake of Antofagasta, Chile, *Geology, 29*, 847–850.

Ichihara, M., and E.E. Brodsky, 2006, A limit on the effect of rectified diffusion in volcanic systems, *Geophys. Res. Lett., 33*, L02316.

Kopf, A.J., 2002, Significance of mud volcanism, *Rev. Geophys., 40*, 1005.

Lemarchand, N., and J.R. Grasso, 2007, Interactions between earthquakes and volcano activity, *Geophys. Res. Lett., 34*, L24303.

Linde, A., and I.S. Sacks, 1998, Triggering of volcanic eruptions, *Nature, 395*, 888–890.

Linde, A.T., I.S. Sacks, M.S.J. Johnston, D.P. Hill, and R.G. Bilham, 1994, Increased pressure from rising bubbles as a mechanism for remotely triggered seismicity, *Nature, 371*, 408–410.

Manga, M., 2007, Did an earthquake trigger the May 2006 eruption of the Lusi mud volcano? *EOS, 88*, 201.

Manga, M., and E. Brodsky, 2006, Seismic triggering of eruptions in the far field: Volcanoes and geysers, *Ann. Rev. Earth Planet. Sci., 34*, 263–291.

Manga, M., M. Brumm, and M.L. Rudolph, 2009, Earthquake triggering of mud volcanoes, *Mar. Petrol. Geol., 26*, 1785–1798.

Marzocchi, W., 2002, Remote seismic influence on large explosive eruptions, *J. Geophys. Res., 107*, 2018.

Mau, S., G. Rehder, I.G. Arroyo, J. Gossler, and E. Suess, 2007, Indications of a link between seismotectonics and CH_4 release from seeps off Costa Rica, *Geochem. Geophys. Geosys., 8*, Q04003.

Mazzini, A., H. Svensen, G. Akhmanov, G. Aloisi, S. Planke, A. Malthe-Sorenssen, and B. Istadi, 2007, Triggering and dynamic evolution of Lusi mud volcano, Indonesia, *Earth Planet. Sci. Lett., 261*, 375–388.

Mellors, R., D. Kilb, A. Aliyev, A. Gasanov, and G. Yetirmishli, 2007, Correlations between earthquakes and large mud volcano eruptions, *J. Geophys. Res., 112*, B04304.

Milkov, A., 2000, Worldwide distribution of submarine mud volcanoes and associated gas hydrates, *Marine Geol., 167*, 29–42.

Newhall, C.G., and S. Self, 1982, The volcanic explosivity index (VEI): An estimate of explosive magnitude for historical volcanism, *J. Geophys. Res., 87*, 1231–1238.

Nishimura, T., 2004, Pressure recovery in magma due to bubble growth, *Geophys. Res. Lett., 31*, L12613, doi:10.1029/2004GL019810.

Pyle, D.M., and D.L. Pyle, 1995, Bubble migration and the initiation of volcanic eruptions, *J. Volcanol. Geotherm. Res., 67*, 227–232.

Rensbergen, P.V., M.D. Batist, J. Klerkx, R. Hus, J. Poort, M. Vanneste, N. Granin, O. Khlystov, and P. Krinitsky, 2002, Sublacustrine mud volcanoes and methane seeps caused by dissociation of gas hydrates in Lake Baikal, *Geology, 30*, 631–634.

Sahagian, D.L., and A.A. Proussevitch (1992) Bubbles in volcanic systems, *Nature, 359*, 485.

Siebert, L., and T. Simkin, 2002, Volcanoes of the world: An illustrated catalog of Holocene volcanoes and their eruptions, Smithsonian Inst. Global Volcanism Digital Inofmration Series, GVP-3, http://www.volcano.si.edu/world.

Sturtevant, B., H. Kanamori, and E.E. Brodsky, 1996, Seismic triggering by rectified diffusion in geothermal systems, *J. Geophys. Res., 101*, 25269–25282.

Sumita, I., and M. Manga, 2008, Suspension rheology under oscillatory shear and its geophysical implications, *Earth Planet. Sci. Lett., 269*, 467–476.

Walter, T.R., and F. Amelung, 2006, Volcano-earthquake interaction at Mauna Loa Volcano, Hawaii, *J. Geophys. Res., 111*, B05204.

Wang, C.-Y., 2007, Liquefaction beyond the near field, *Seismol. Res. Lett., 78*, 512–517.

Watt, S.F.L., D.M. Pyle, and T.A. Mather, 2009, The influence of great earthquakes on volcanic eruption rate along with Chilean subduction zone, *Earth Planet. Sci. Lett., 277*, 39–407.

Yang, T.F., C.C. Fu, V. Walia, C.-H. Chen, L.-L. Chyi, T.-K. Liu, S.-R. Song, M. Lee, C.-W. Lin, and C.C. Lin, 2006, Seismo-geochemical variations in SW Taiwan: Multi-parameter automatic gas monitoring results, *PAGEOPH, 163*, 693–709.

Youd, T.L., J.H. Steidl, and R.L. Nigbor, 2004, Lessons learned and need for instrumented liquefaction sites, *Soil Dyn. Earthq Eng., 24*, 639–646.

Chapter 4
Increased Stream Discharge

Contents

4.1 Introduction

Changes in stream discharge after earthquakes are among the most interesting hydrologic responses partly because they can be directly observed, are persistent, and may be large enough to be visually compelling. These changes have also been quantitatively documented for a long time – perhaps longer than any other hydrological responses. For example, extensive networks of stream gauges in the western United States were established by the US Geological Survey in the early twentieth century, and long and continuous gauging measurements have been collected. Some of the changes following earthquakes reflect changes in surface hydrology, including

C.-Y. Wang, M. Manga, *Earthquakes and Water*, Lecture Notes in Earth Sciences 114, DOI 10.1007/978-3-642-00810-8_4, © Springer-Verlag Berlin Heidelberg 2010

instant waterfalls created by the earthquake faulting, decreases of downstream discharge caused by the damming of mountain valleys by landslides and rockfalls, and increases of discharge in regions of high relief caused by the avalanche of large quantities of snow to lower elevations, increasing the supply of melt water. Such changes only redistribute the surface discharge budget, with excess and deficit flow compensating each other over a period of days. More interesting and much less well understood is a type of discharge change that follows earthquakes and persists for an extended duration (commonly several weeks to months) but has no obvious source. A dramatic example of this type was the streamflow increase after the 1989 M_w 6.9 Loma Prieta earthquake (Rojstaczer and Wolf, 1992) which occurred during a prolonged drought in California when the stream discharge before the earthquake was very low (Fig. 4.1a). After the earthquake, many streams within about 50 km of the epicenter showed increased discharge by a factor of 4 to 24; the total excess discharge was estimated to be 0.01 km^3 (Rojstaczer et al., 1995).

This total amount turns out to be notably smaller in comparison with some estimated excess discharges following other large earthquakes, i.e., 0.7–0.8 km^3 after the *M* 7.5 Chi-Chi earthquake (Wang et al., 2004a), 0.5 km^3 after the *M* 7.5 Hebgen Lake earthquake (Muir-Wood and King, 1993), and 0.3 km^3 after the *M* 7.3 Borah Peak earthquake (Muir-Wood and King, 1993). This latter type of discharge response is the focus of the present chapter.

In the following, we first discuss some general characteristics of the observations, which will lead to a discussion on how to estimate the amount of 'excess' stream discharge after earthquakes. We then discuss several mechanisms that have

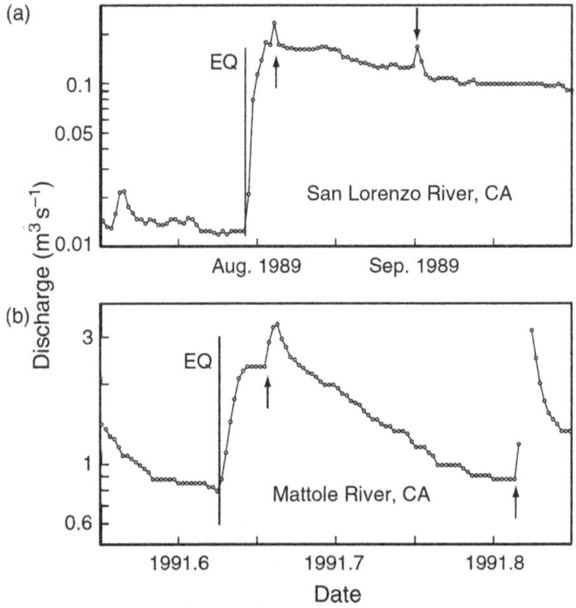

Fig. 4.1 Hydrographs of (**a**) San Lorenzo River, CA, and (**b**) Mattole River, CA, showing postseismic response to the 1989 M_w6.9 Loma Prieta earthquake and 1991 M_s6.2 Honeydew earthquake, respectively. The vertical line labeled EQ indicates the time of earthquake. Small arrows show the time of rainfall events. Figure made with US Geological Survey stream gauge data (From Manga and Wang, 2007)

been proposed to explain the stream responses to earthquakes. Following this, we discuss how these mechanisms perform when tested against observations. Finally, we discuss the occurrence of streamflow responses in special geologic settings, such as geothermal areas and subduction zones.

4.2 Observations

Most studies of streamflow increases have focused on the response of a set of streams to a given earthquake (e.g., Rojstaczer and Wolf, 1992; Muir-Wood and King, 1993; Sato et al., 2000; Montgomery et al., 2003; Wang et al., 2004a). As an example, Fig. 4.2 shows the stream responses to the 1959 M7.5 Hebgen Lake earthquake (Muir-Wood and King, 1993).

As far as can be determined from the stream-gauge records (e.g., Figs. 4.1 and 4.2), the onset of streamflow increase is coseismic. The increase can, however, continue for a few days to reach a maximum, and then gradually declines to reach the pre-earthquake level after several months. Also noticeable in Fig. 4.1 is the sudden

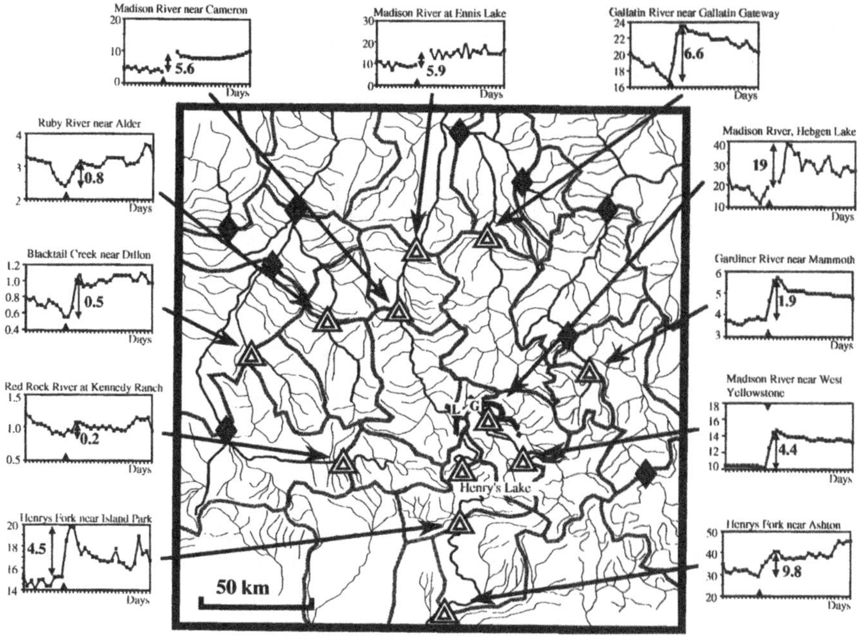

Fig. 4.2 River basins (outlined by *thick lines*) and daily flow data before and after the 1959 Hebgen Lake earthquake. *Double triangles*: river gauges with quantified increases of flow; *diamonds*: no discernable changes; *L*: landslide dam. In each of the small figures, the *dots* are daily average discharge in m^3/s, and the time of the earthquake is indicated by a *small triangle*. Numbers associated with double-ended *arrows* show the estimated peak excess flow in m^3/s (Reproduced from Muir-Wood and King, 1993)

increase in streamflow in response to local precipitation. Precipitation can easily obscure the earthquake-induced streamflow response when it occurs at the time of an earthquake and makes the analysis of the latter difficult or impossible. On the other hand, if no precipitation occurs or if the precipitation-induced streamflow is too small to obscure the earthquake-induced response, the hydrograph may yield valuable information such as the amount of excess discharge 'produced' by the earthquake, as explained in the next section.

4.3 Characteristics of Increased Discharge

The characteristics of the increased stream discharge following an earthquake can be expressed by a postseismic baseflow recession or its characteristic time. The amount of excess discharge following earthquakes is another interesting characteristic. In this section we show with an example how these parameters may be estimated from the streamflow data.

Seventeen stream gauges were installed on three stream systems near the Chi-Chi earthquake epicenter (Fig. 4.3). During and after the Chi-Chi earthquake, many of these gauges registered large increases in stream discharge (WRB, 2000). These records provide an excellent opportunity for testing the hypotheses proposed to explain the observed increases, and is used here as an example to illustrate how the postseismic recession constant, the characteristic time and the amount of streamflow increase are estimated.

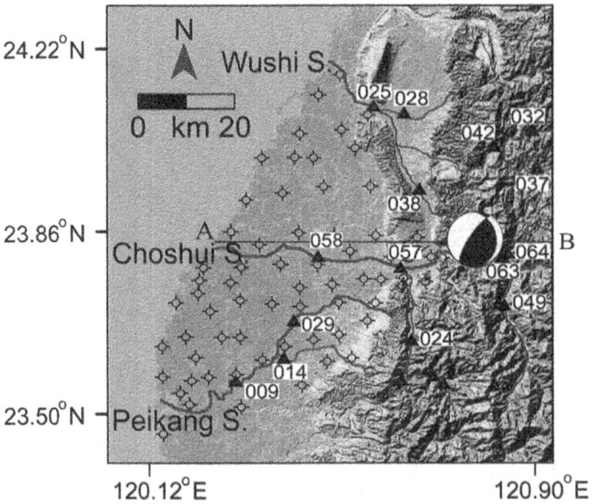

Fig. 4.3 Map shows the three stream systems and stream gauges (in *black triangles*) installed along these systems. Numbers are gauge numbers for identification. The epicenter and focal mechanism of the Chi-Chi earthquake are indicated. Choshui alluvial fan is on west side and foothills are on east side. *Open circles* with *crosses* show well locations. Stream systems are labeled as Choshui S. for Choshui Stream, etc. Tributaries are not labeled because of space. AB marks the location of geologic cross-section shown in Fig. 5.3

4.3.1 Recession Analysis

Figure 4.4 shows the hydrograph of a stream (at gauge #32) in the mountains before and after the Chi-Chi earthquake. The figure also shows the precipitation record from a nearby station. Since there was little precipitation in the area after the Chi-Chi earthquake, the hydrograph may be used to analyze the baseflow, i.e., the component of stream discharge from groundwater seeping into the stream. The technique is widely known in hydrology as baseflow recession analysis and is used to gain an understanding of the basin-scale discharge processes that make up the baseflow. The figure shows that, on a *ln Q* versus time diagram, where *Q* is the discharge of the stream, and after a sufficient lapse of time following the earthquake, the data points fall closely along a straight line (Fig. 4.4), i.e.,

$$ln\,Q = a - c\,t \tag{4.1}$$

where *a* and *c* are the empirical constants for the linear fit. A minus sign is placed in front of *c* so that *c* itself is positive; *c* is known as the recession constant; its inverse, i.e., $\tau \equiv 1/c$, is the time scale that characterizes the rate at which the groundwater discharge decreases. Table 4.1 lists the values for *c* and τ for a number of streams from the gauge records following the Chi-Chi earthquake.

Although the values of *c* and τ as defined above are entirely empirical, they are closely related to the geometrical and physical nature of the aquifer, as made clear by the following simplified model. Since the aquifers are approximately horizontal with a length scale much greater than their thickness, we may assume, to a first-order approximation, a one-dimensional aquifer that extends from one end of the aquifer at $x = 0$ to the other end at $x = L$, as shown in Fig. 4.5a. The time-dependent discharge from this aquifer may be simulated by solving the following flow equation (B.25), under appropriate boundary and initial conditions:

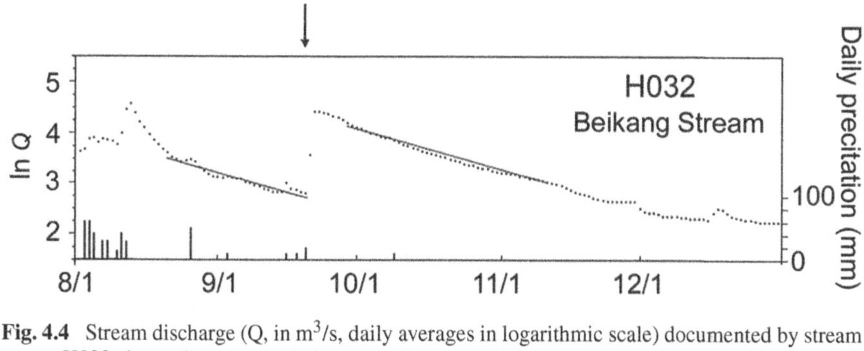

Fig. 4.4 Stream discharge (Q, in m^3/s, daily averages in logarithmic scale) documented by stream gauge H032, located on a stream in mountains (see Fig. 4.3b for location). Note the surge in discharge right after the 1999 Chi-Chi earthquake as indicated by the downward pointed *arrow*. Small bars at *bottom* show precipitation records from a nearby station. Notice that there was little precipitation in the area after the Chi-Chi earthquake . Linear segments of data before and after the earthquake fitted by *straight lines* show similar slopes (Modified from Wang et al., 2004a)

Table 4.1 Recession constant c and characteristic time τ from recession analysis of some stream gauge data after the Chi-Chi earthquake

Streams	Stream gauges	c (s^{-1})	τ (s)
Wushi system			
Wushi	H025	5.2×10^{-7}	1.9×10^{6}
Beikang	H032	5.0×10^{-7}	2.0×10^{6}
Nankang	H037	5.0×10^{-7}	2.0×10^{6}
Wushi	H042	4.7×10^{-7}	2.1×10^{6}
Choshui system			
Choshui	H057	3.5×10^{-7}	2.9×10^{6}
Choshui	H058	6.7×10^{-7}	1.5×10^{6}
Choshui	H063	5.0×10^{-7}	2.0×10^{6}

$$S_s \frac{\partial h}{\partial t} = K \frac{\partial^2 h}{\partial x^2} + A \tag{4.2}$$

where h is the *excess* hydraulic head above the background value, K is the *horizontal* hydraulic conductivity, S_s is the specific storage, and A is the rate of recharge to the aquifer per unit volume. Even though this model is highly simplified, several studies (e.g., Roeloffs, 1998; Manga, 2001; Manga et al., 2003; Brodsky et al., 2003; Wang

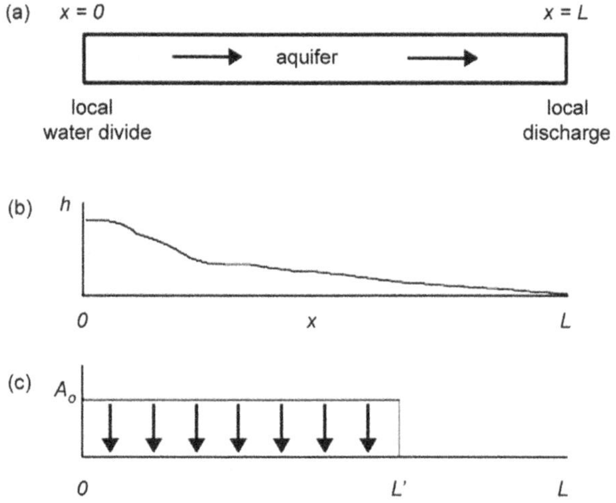

Fig. 4.5 (**a**) Schematic drawing to show the geometry of the model aquifer and its boundaries: a local water divide is located at $x = 0$, and a local discharge is located at $x = L$. (**b**) Schematic drawing to show the boundary conditions used in solving Eq. (4.2): head 'h' is zero at the local discharge ($x = L$) and the gradient of head is zero at the local water divide ($x = 0$). (**c**) Schematic drawing to show the assumed recharge used in solving (4.2): at $t = 0$, recharge is H_o if $x \leq L'$ and zero if $x > L'$

et al., 2004a) have demonstrated that the procedure is useful for characterizing the catchment-scale response of hydrological systems to earthquakes. Equation (4.2) is also the linearized form of the differential equation that governs the groundwater level in unconfined aquifers, but with S_s replaced by S_y/b where S_y is the specific yield and b the saturated thickness of the unconfined aquifer. Because these equations are linear, the head change due to the earthquake may be superimposed on the background hydraulic head.

For boundary conditions, we adopt a no-flow boundary condition at $x = 0$ (i.e., a local water divide) and $h = 0$ at $x = L$ (i.e., a local discharge to a stream) (Fig. 4.5b). Taking the background head as the reference value, we have the initial condition $h = 0$ at $t = 0$. Under these conditions, solution to the flow equation (4.2), as given later, shows that c and τ are related to two important aquifer properties, hydraulic diffusivity $D = K/S_s$ and length L, by

$$c \equiv -\frac{\partial \log Q}{\partial t} \approx \frac{\pi^2 D}{4L^2}, \tag{4.3}$$

after a sufficient lapse of time following the earthquake. Thus

$$\tau \equiv \frac{1}{c} \approx \frac{4L^2}{\pi^2 D}. \tag{4.4}$$

The recession constant and the characteristic time of an aquifer may also be estimated from the postseismic decline of groundwater level in wells, as shown in the next chapter.

4.3.2 Estimate Excess Discharge

The solution of Eq. (4.2) will depend on the temporal and spatial distribution of the recharge function $A(x,t)$. Since the duration of water release during the earthquake is very brief in comparison with the duration for the postseismic evolution of $h(x,t)$, we may simplify the problem by assuming it to be instantaneous; i.e.,

$$A(x,t) = A_o(x)\,\delta\,(t = 0) \tag{4.5}$$

where $A_o(x)$ is the amount of earthquake-induced recharge to the aquifer per unit volume, and $\delta(t = 0)$ is a delta function that equals 1 when $t = 0$ and equals zero when $t > 0$. If we further simplify the problem by assuming $H_o(x) = \int_0^t A_o(x)\delta(t)\,dt = H_o$ for $x \leq L'$ and zero for $x > L'$ (Fig. 4.5c), we have the solution for Eq. (4.2) (Appendix B):

$$h(x,t) = \frac{4H_o}{\pi S_s} \sum_{n=1}^{\infty} \frac{1}{n} \sin\frac{n\pi L'}{2L} \cos\frac{n\pi x}{2L} \exp\left[-n^2\frac{t}{\tau}\right]. \tag{4.6}$$

The excess discharge to the stream is given by

$$Q_{ex} = -KF \frac{\partial h}{\partial x} \tag{4.7}$$

evaluated at $x = L$, where F is the cross-section surface area of the discharging aquifer. Thus we have the excess discharge of the aquifer at $x = L$:

$$Q_{ex} = \frac{2DVH_o}{L^2 (L'/L)} \sum_{r=1}^{\infty} (-1)^{r-1} \sin\left[\frac{(2r-1)\pi L'}{2L}\right] \exp\left[-\frac{(2r-1)^2 \pi^2 D}{4L^2} t\right], \tag{4.8}$$

where $V = FL$' is the volume of the aquifer between $x = 0$ and L' (Fig. 4.5c), and VH_o is the total amount of excess water recharging the aquifer.

The amount of earthquake-induced excess discharge may be estimated by 'fitting' the streamflow data with the above equation. Figure 4.6 shows such a 'fit' for the streamflow data from stream gauge H032 (Fig. 4.4), adjusted to a reference of $Q_{ex} = 0$ before the Chi-Chi earthquake. An excellent fit is obtained with $D/L^2 = 2.4 \times 10^{-7}\ s^{-1}$ (Table 4.1) and $L'/L = 0.8$. The latter is consistent with the fact that the stream gauge H032 is located in the mountains where the width of the flood plain is relatively small.

With the values of D/L^2 calculated from c or τ (Table 4.1) using Eq. (4.3), and model fitting using Eq. (4.8), we obtain the amount of the postseismic excess water, VH_0, for several stream gauges (Table 4.2). By summing the excess discharges in the Choshui Stream and in the Wushi Stream, we obtain a total amount of excess discharge of 0.7–0.8 km^3 after the Chi-Chi earthquake.

Returning to the physical meaning of the postseismic baseflow recession we differentiate Eq. (4.8) with respect to t and find that, for sufficiently long time after the

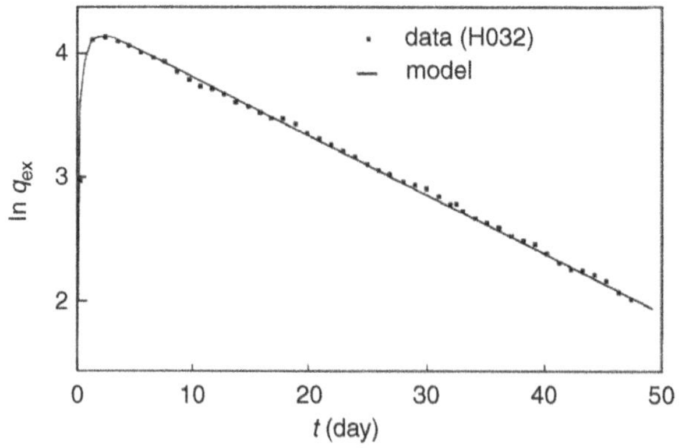

Fig. 4.6 Diagram comparing the excess postseismic discharge (Q_{ex}, in m^3/s) predicted by model (*curve*) with daily averaged data (*dots*) from stream gauge H032. See text for details (Reproduced from Wang et al., 2004a)

Table 4.2 Estimated postseismic excess discharge in some streams after the Chi-Chi earthquake

Streams	Stream gages	VH_0(km^3)
Wushi	*River*	
Wushi	H025	0.21
Beikang	H032	0.14
Nankang	H037	0.10
Wushi	H042	0.23
Choshui	*River*	
Choshui	H057	0.56
Choshui	H058	0.55
Choshui	H063	0.44

earthquake such that $t \geq \frac{4L^2}{\pi^2 D}$, the stream discharge, Q_{ex}, decreases exponentially with increasing time as $\exp\left(-\frac{\pi^2 D}{4L^2}t\right)$. Thus we have

$$c \equiv -\frac{\partial \log Q_{ex}}{\partial t} \approx \frac{\pi^2 D}{4L^2},$$

as given in Eq. (4.3).

For the example shown in Fig. 4.6, the peak postseismic discharge occurs within two days of the earthquake. Figure 4.7 shows an example in which the peak discharge is reached 9 to 10 days later, though discharge begins to increase coseismically. This example is for Sespe Creek, CA, responding to the 1952 *M* 7.5 Kern County earthquake located 63 km away from the center of the drainage basin. Again, the model for excess discharge given by Eq. (4.8), shown in the solid curve

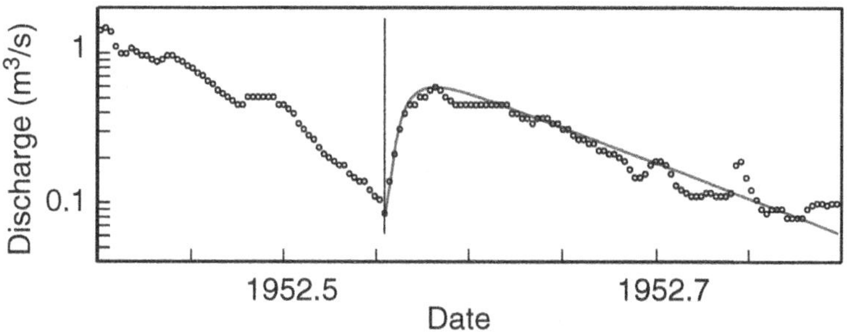

Fig. 4.7 Response of Sespe Creek, CA to the 1952 *M* 7.5 Kern County earthquake. Daily discharge measurements collected and provided by the US Geological Survey are shown with *circles*. *Curve* is solution to Eq. (4.8) for the excess flow with $L'/L = 0.4$ added to the baseflow, to recover the entire hydrograph. *Vertical line* shows the time of the earthquake. There was no precipitation during the entire time interval shown in this graph (Modified from Manga et al., 2003)

in Fig. 4.7, fits the observed postseismic discharge very well (baseflow has been added back to the calculated excess discharge).

4.4 Proposed Mechanisms

In the absence of recent precipitation or snowmelt, a change in stream discharge implies either a change in the hydraulic gradient created by a new source of head or a change in the permeability along the flow path, or both (see Eq. (4.7)). Several mechanisms have been proposed to explain the changes in streamflow following earthquakes: (1) expulsion of deep crustal fluids resulting from coseismic elastic strain (e.g., Muir-Wood and King, 1993), (2) changes in near-surface permeability (Briggs, 1991; Rojstaczer and Wolf, 1992; Rojstaczer et al., 1995; Tokunaga, 1999; Sato et al., 2000), and (3) consolidation or liquefaction of near-surface deposits (Manga, 2001; Manga et al., 2003; Montgomery et al., 2003). The differences between these different hypotheses are nontrivial because they imply different hydrologic processes during and after earthquakes, and have implications for the nature of groundwater flow paths. In the following we summarize the basic elements and some problems with each hypothesis and then consider the streamflow response to the 1999 Chi-Chi earthquake in detail as an example to evaluate the proposed hypotheses.

4.4.1 Coseismic Elastic Strain

Muir-Wood and King (1993) applied the coseismic elastic strain model, proposed by Wakita (1975) for explaining coseismic groundwater-level changes, to explain the increased stream discharge after the 1959 *M* 7.5 Hebgen Lake earthquake and the 1983 *M* 7.3 Borah Peak earthquake . They noted that changes in the static elastic strain in the crust produced by earthquake faulting cause rocks to dilate or contract and thus saturated cracks in rocks to open or close, resulting in a decrease or increase in the groundwater discharge into streams (Fig. 4.8). In order to account for the extra water in the increased streamflow by the groundwater expelled through coseismic elastic strain, a very large volume of the crust must be involved (Rojstaczer et al., 1995); thus the model of Muir-Wood and King (1993) implicitly requires that the streamflow increases are related to mid-crustal processes.

4.4.2 Enhanced Permeability

Briggs (1991) and Rojstaczer and Wolf (1992; also Rojstaczer et al., 1995) proposed a model of enhanced permeability of the shallow crust resulting from seismically induced cracks and fractures to explain the increased stream discharge and the changes in the ionic concentration of stream water following the 1989 Loma Prieta

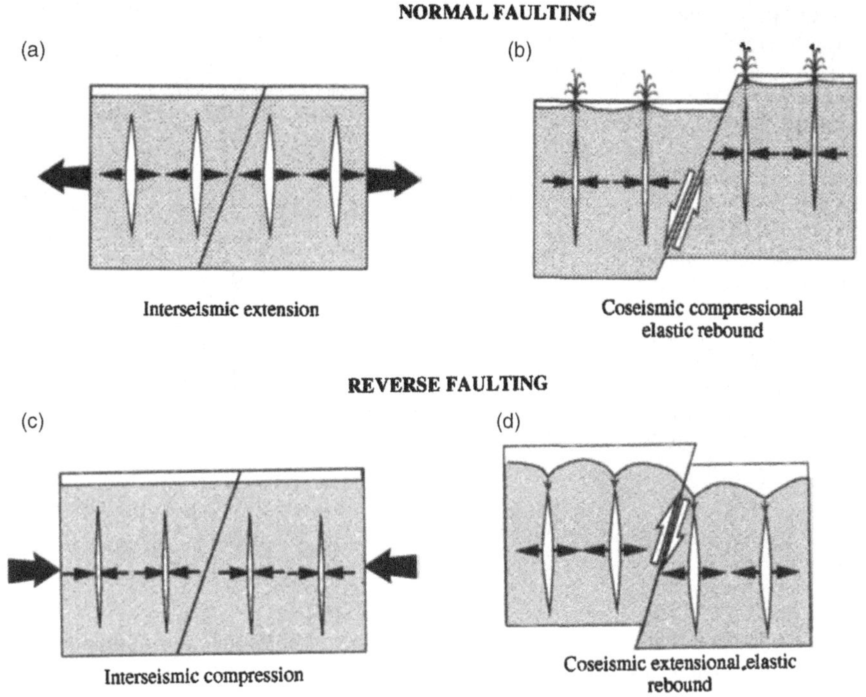

Fig. 4.8 Simplified model for the interseismic accumulation and coseismic release of strain in extensional and compressional tectonic environments. (**a**) For extensional faulting, the interseismic period is associated with crack opening and increase of effective porosity. (**b**) At the time of the earthquake, cracks close and water is expelled. (**c**) For compressional faulting, the interseismic period is associated with crack closure and the expulsion of water. (**d**) At the time of the earthquake, cracks will open and water will be drawn in. In the case of normal faulting, water can be expelled on to the surface at the time of an earthquake and thus immediately affect river flow (shown schematically as surface fountains). For reverse faulting, cracks must be filled from the water table, a slower process that may not be observed in river flow rates (Reproduced from Muir-Wood and King, 1993)

earthquake in California. Similar models were applied to the 1995 Kobe earthquake in Japan to explain the observed hydrological changes (Tokunaga, 1999; Sato et al., 2000). Permeability enhancement was also invoked to explain the increased electrical conductivity of water discharged after an earthquake (Chamoille et al., 2005). Furthermore, Elkhoury et al., (2006) found that earthquakes in southern California caused phase shifts in the water-level response to tidal strain, and interpreted these to be due to increased permeability created by seismic waves.

4.4.3 Coseimic Consolidation and Liquefaction

In view that undrained consolidation of saturated soils can increase pore pressure in saturated soils and cause liquefaction (Chap. 2), Manga (2001; also Manga et al.,

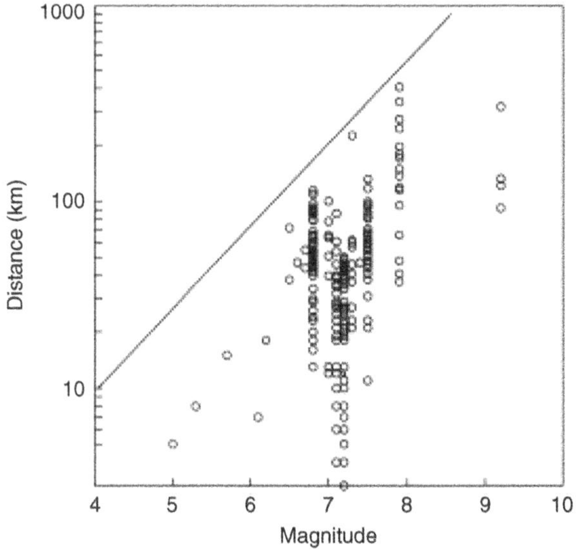

Fig. 4.9 Relationship between earthquake magnitude and distance from the epicenter of streams that exhibited clear and persistent increases in discharge. The *solid line* is an empirical upper bound for the maximum distance over which shallow liquefaction has been observed (Wang, 2007): $M = 2.0 + 2.1 \log R$, where R is in kilometer. Data shown based on the compilation in Montgomery and Manga (2003) with additional data (From Wang et al., 2005)

2003) suggested that coseismic liquefaction of loose sediments on floodplains may provide the water for the increases in stream discharge following earthquakes. Figure 4.9 shows the relationship between earthquake magnitude and distance between the epicenter and the center of the gauged basin for streams that responded to earthquakes. Also shown for reference is the liquefaction limit, i.e., the maximum distance over which liquefaction has been reported (from Fig. 2.6).

The coincidence of the liquefaction limit and the limit for the occurrence of streamflow increase is suggestive, but does not require, that the two processes may be causally related. Montgomery et al. (2003) searched for a field association between liquefaction and increased streamflow after the 2001 *M* 6.8 Nisqually, WA, earthquake, but found none.

4.5 Debate About Mechanisms

4.5.1 Geochemical and Temperature Constraints

If the observed excess stream discharge following earthquakes was caused by the expulsion of water from the mid-crust after the earthquake, as required by the elastic strain model, there would have been unusual geochemical changes in the stream water due to different concentrations of ions in the mid-crustal fluids, and there would also have been significant increases in the stream water temperature. However, neither of these predictions were verified. On the contrary, after the 1989 Loma Prieta earthquake the ratios of various ionic constituents in the stream water remained nearly the same as those before the earthquake, even though the concentrations of various ions did increase (Rojstaczer et al. 1995). Furthermore, the temperature of the stream water decreased, instead of increasing, supporting a shallow origin of the excess water. The constant ionic ratios and the decreased water temperature both contradict the prediction of the elastic strain model, but can be easily explained by an increased baseflow contribution due to an increased permeability in shallow aquifers or by an increase in head created by local liquefaction.

4.5.2 Constraints from Multiple Earthquakes

Although most studies on earthquake-induced streamflow changes have focused on the response of multiple streams to a single earthquake, some studies have focused on the response of a single stream to multiple earthquakes (Leonardi et al., 1990; King et al., 1994; Manga et al., 2003; Manga and Rowland, 2009). Such studies were made possible by the availability of long, continuous gauging measurements. Manga et al. (2003) took advantage of data collected by the USGS (1928-present) and the relatively high rate of seismicity in southern California to characterize the response of Sespe Creek, California, to several earthquakes. Figure 4.10 shows the location of the stream together with the epicenters and the focal mechanisms of several large

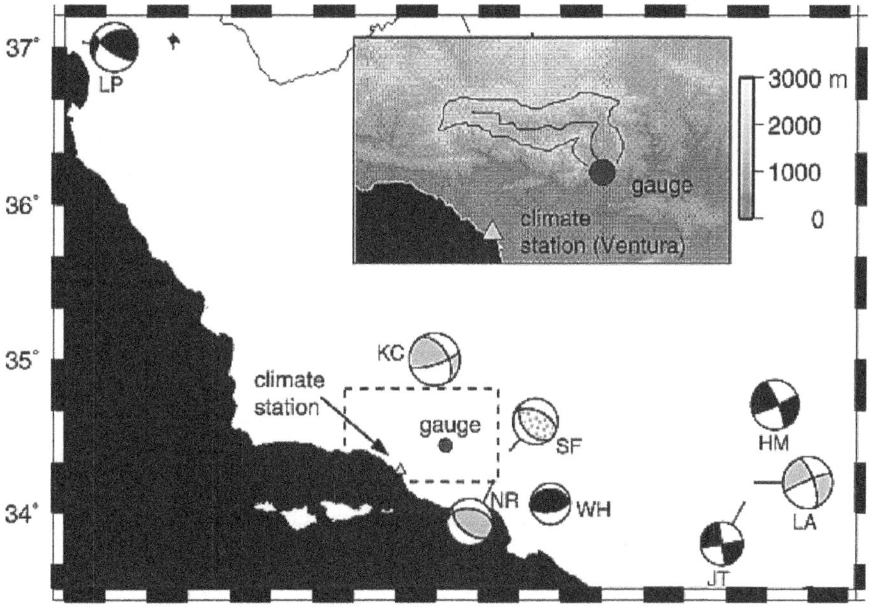

Fig. 4.10 Map showing the Sespe Creek basin in southern California and the location of stream gauge, together with the epicenters and focal mechanisms of several large earthquakes (*grey* – streamflow increase, *grey dots* – possible increase, *black* – no change). Inset shows the region in the *dashed box* (From Manga et al., 2003)

earthquakes. Manga et al. (2003) found that the streamflow responses in the Sespe Creek basin always increased regardless of whether the earthquake-induced static strain in the basin was contraction or expansion. One example is shown in Fig. 4.7. This finding ruled out the static strain hypothesis as a viable mechanism, at least for this basin.

4.5.3 Constraints from Recession Analysis

Manga (2001) analyzed the hydrographs of streams near the epicenter of the 1989 M_w6.9 Loma Prieta earthquake, before and after the earthquake (Fig. 4.11). The postseismic baseflow recession is shown by the bold sloping line in Fig. 4.11a. Figure 4.11b shows the constant c determined from the hydrographs; no significant change in baseflow recession was found before and after the earthquake, even though discharge increased by an order of magnitude after the earthquake.

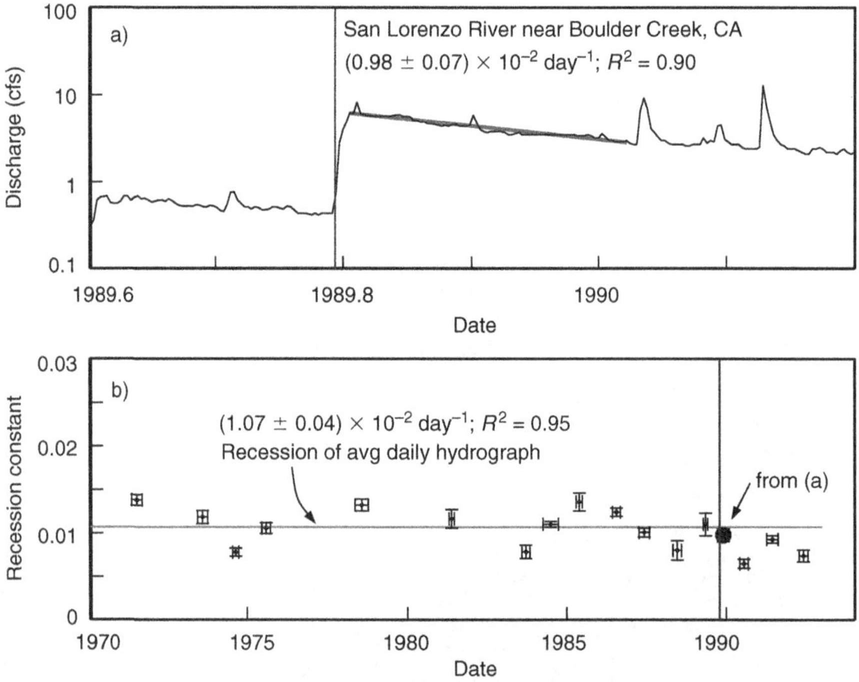

Fig. 4.11 (**a**) Hydrograph of the San Lorenzo River, CA, showing postseismic response to the 1989 M 6.9 Loma Prieta earthquake. The *vertical line* indicates the time of the earthquake. The postseismic period of baseflow recession is shown by the *bold sloping line*. (**b**) The baseflow recession constant for periods of baseflow before and after the earthquake shows that even though discharge increased by an order of magnitude after the earth quake there was no significant change in baseflow recession. Figure made with US Geological Survey stream gauge data (Reproduced from Manga, 2001)

Referring to Eq. (4.3), an unchanged baseflow recession implies that aquifer permeability did not change after the earthquake, given that the aquifer dimension and flow pathways are not likely to change during the earthquake. The above result was substantiated by later studies in other areas (Montgomery et al., 2003; Wang et al., 2004a) and during other earthquakes (Manga et al., 2003). This finding argues against an earthquake-enhanced permeability and thus *at first glance* appears to pose a problem for the enhanced permeability model.

4.5.4 Constraints from Multiple Stream Gauges

As mentioned earlier (and shown in Fig. 4.3b), an extensive network of stream gauges is installed on three stream systems in central Taiwan. Among the three systems, two (Choshui Stream and the Wushi Stream) have many tributaries in the mountains, but the third (Peikang Stream) does not but instead originates on the western edge of the frontal thrust (Fig. 4.3). After the Chi-Chi earthquake, all the tributaries in the mountains showed large postseismic streamflow increases (Table 4.2). On the alluvial fan, the Choshui Stream and the Wushi Stream, both with tributaries in the mountains, also showed large increases in streamflow, but the amount of increase in the proximal area of the Choshui alluvial fan was the same as that in the distal area of the fan, suggesting that there was little contribution of water from the sediments in the fan. In contrast, the Peikang Stream system, which does not have tributaries in the mountainous area, did not show any noticeable postseismic streamflow increases. We thus conclude that the excess discharge after the Chi-Chi earthquake originated from the mountains where little loose sediments exist, and any contribution from coseismic consolidation and liquefaction in the floodplain (alluvial fan) must have been insignificant. This result appears to rule out the consolidation hypothesis outlined in Sect. 4.4.3.

4.5.5 Role of Anisotropic Permeability

As discussed above, none of the mechanisms proposed so far can explain all the existing observations. The static elastic strain model is contradicted by temperature and composition data for the stream water after the earthquake. Furthermore, it fails to explain the persistent increase in streamflow at given sites in response to multiple earthquakes of different mechanisms and different fault orientations. The model of consolidation and liquefaction cannot explain why the increased discharge after the Chi-Chi earthquake was derived mainly from the mountains with virtually no sediments, but *not* from the flood plain where liquefaction was widespread. Furthermore, field searches failed to find evidence for an association between the occurrence of liquefaction and streamflow changes. Finally, the enhanced-permeability model appears to be contradicted by the results of recession analysis, which suggests that there was apparently no change in permeability after earthquakes.

Wang et al. (2004a) suggest that the apparent contradiction between the enhanced permeability model and the recession analysis may be resolved by recognizing that the permeability of the shallow crust and its changes during earthquakes may be anisotropic. The foothills of the Taiwan mountains, for example, consist of alternating layers of sandstone and shale. Thus, before the Chi-Chi earthquake, the vertical permeability, which is controlled by the most impervious member of the sedimentary layers, i.e., shales, is very much smaller than that in the horizontal direction, which is controlled by the most permeable member of the layers, i.e., sandstones and gravels (Appendix B). Thus groundwater flow before the earthquake was sub-horizontal (Fig. 4.5a), along the sand and gravel beds. After the Chi-Chi earthquake, numerous subvertical tensile cracks appeared in the hanging wall of the thrust fault that cut across the sedimentary beds (Angelier et al., 2000; Lee et al., 2000, 2002b). Many wells in the foothills above the thrust fault showed a significant drop in water level and a tunnel beneath the foothills experienced a sudden downpour right after the earthquake (Lin, 2000; Yan, 2001; Chia et al., 2001; Wang et al., 2001). Thus the Chi-Chi earthquake may have greatly increased the vertical permeability of the sedimentary layers that allowed rapid downward draining of groundwater to recharge aquifers beneath the foothills (Fig. 4.5c). The horizontal permeability of the aquifer, which was already high before the earthquake, may not be significantly affected by the vertical cracks formed during the earthquake. Since the baseflow recession is controlled by the horizontal permeability, it is easy to understand why the baseflow recession constant before and after the earthquake did not change. Finally, as Fig. 4.6 shows, the predicted streamflow from this model of anisotropic permeability of aquifers and its earthquake-induced change agrees very well with observation.

Lowering of the water level in mountainous terrains was also reported near the epicenters of the Loma Prieta earthquake in California (Rojstaczer et al., 1995) and of the Kobe earthquake in Japan (Tokunaga, 1999) after the respective earthquakes. Both these earthquakes occurred in a transform tectonic setting, in contrast with the convergent tectonic setting of the Chi-Chi earthquake. Thus the process of earthquake-induced release of water from mountains may be insensitive to different tectonic settings. The variability in the volume of excess flow in different regions following different earthquakes, i.e., 0.8 km^3 after the M 7.5 Chi-Chi (Taiwan) earthquake versus 0.01 km^3 after the M 7.1 Loma Prieta (California) earthquake, may reflect the differences in regional recharge, i.e., a precipitation of $\sim$2.5 m/yr in Taiwan vs. $<$0.5 m/yr in the Santa Cruz (California) area. Other factors such as differences in topography (affecting hydraulic gradient) may also change the volume and the rate of excess flow.

The vertical permeability may readily return to its pre-earthquake state after a large earthquake, owing to active biogeochemical processes; thus the vertical recharge of groundwater may again be impeded. Postseismic recovery of permeability in Iceland took about 2 years (Claesson et al., 2007; Chap. 6). Similar repose times were documented for earthquake triggering of mud volcanoes (Mellors et al.,

2007; Manga et al., 2009; Chap. 3). Thus the time scale for the postseismic recovery of permeability is much shorter than the recurrence time of large earthquakes. Thus the mountain slopes will be recharged with groundwater, only to be released once more during the next large earthquake. Based on the recurrence rate of large earthquakes in Taiwan, Wang et al. (2004a) estimated a residence time of 10^4 yr for the groundwater in the foothills. Thus the foothills of active mountain belts, such as Taiwan, may be repeatedly flushed by meteoric water in geologically short time, which may have significant implication on the geochemical evolution of the rocks and groundwater in the mountains.

4.6 Streamflow Increase in Hydrothermal Areas

Within 15 min. of the 2003 M6.5 San Simeon earthquake in central California, two stream gauges registered increased stream discharge, one along the Salinas River near the town of Paso Robles and the other along the Lopez Creek near the town of Arroyo Grande (Fig. 4.12), both known for their hot springs. As explained below, these streamflow increases can be explained by the enhanced permeability model explained earlier, but were apparently driven by the excess pore pressure in a geothermal reservoir, entirely different from those discussed earlier which were driven by gravitational potential.

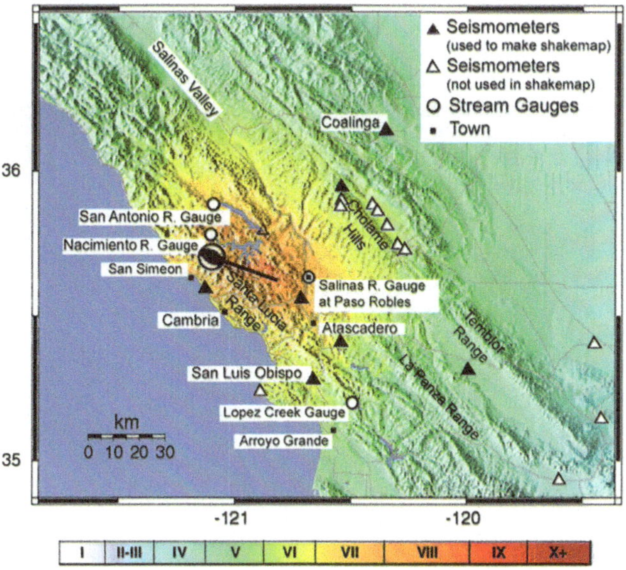

Fig. 4.12 Map showing the instrumental intensity of ground shaking in San Simeon earthquake (modified from Hardeback et al., 2004). Focal mechanism of the earthquake is taken from Harvard CMT Catalog. *Bold line* shows rupture extent. Salinas River flows NW through town of Paso Robles and Salinas Valley

Some background information about the local geology and climate may be required to better understand the different responses of these streams. Active tectonics since the late Tertiary has repeatedly faulted and uplifted the Coast Ranges of California. The climate of the area is semiarid. Most of the annual 250–330 mm precipitation occurs during the winter. The Salinas River, with a flood plain ~100 m wide, runs NW through the Paso Robles Basin. A basin-wide decline of the groundwater level has occurred during the past several decades, caused by a growing population and increased urbanization and agriculture. As a result, the streambed is usually dry except during rainy season, and it was dry before the San Simeon earthquake. The towns of Paso Robles and Arroyo Grande are well known for their hot springs (~40°C). Drilling at Paso Robles encountered the hydrothermal reservoir at a depth of ~100 m. On the other hand, no hot springs are known in the nearby valleys of the San Antonio River or the Nacimiento River.

The epicenter of the San Simeon earthquake occurred 11 km NE of the town of San Simeon and 39 km WNW of Paso Robles. Rupture during the earthquake shows a strong ESE directivity (Fig. 4.12). Following the earthquake, four new hot springs appeared on the two sides of the Salinas River (Fig. 4.13) near the town of Paso Robles. These new hot springs occurred along a straight line striking WNW, parallel to the earthquake rupture (Fig. 4.12) and crossing the Salinas River ~1 km upstream of the stream gauge. The well-head pressure at an established hot spring well (Fig. 4.13) in Paso Robles was steady before the earthquake, but decreased from at 0.33 MPa to ~0.2 MPa within 2 days after the earthquake.

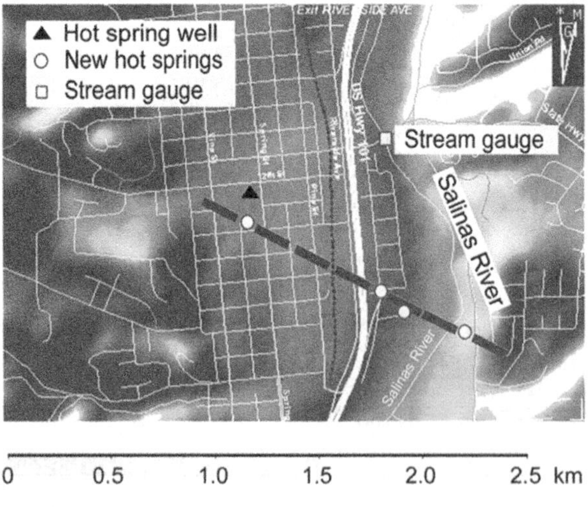

Fig. 4.13 Map of Paso Robles showing locations of stream gauge and four new hot springs after the earthquake. Note the four new hot springs lie along a straight line that is parallel to the fault direction shown in Fig. 4.12. This line intersects Salinas River ~1 km upstream of the stream gauge. The location of a hot spring well, established long before the earthquake, is marked by a black triangle

Recession analysis (Sect. 4.3.1) of the postseismic stream discharge in the Salinas River and the Lopez Creek yields a small characteristic time (~40 min.), suggesting that the sources of the extra water were in close proximity. However, there was no surface water source in the Paso Robles Basin and any surface water would have to be supplied from distant mountains, which is contradicted by the short timescale required by the data. Thus, as suggested by the appearance of new hot springs in the area (Fig. 4.13), the source for the increased discharge in the two streams was likely a subsurface hydrothermal reservoir. An ideal test of this hypothesis would have been a chemical analysis of the increased flow. Unfortunately, the duration of the extra discharge was short and precipitation in the area started one day after the earthquake, which made such analysis unattainable.

The excess discharge based on this model (Fig. 4.14) may be calculated by using Eq. (4.8). The results, shown in the solid curve in Fig. 4.15, fit the observed postseismic discharges (Wang et al., 2004b). Estimated excess discharge after this earthquake ranged from 10^2 to 10^3 m^3, much smaller than the examples mentioned in the Introduction.

Abrupt increases in streamflow and hotspring discharge after earthquakes were also reported in other hydrothermal areas such as in the Long Valley, California (Sorey and Clark, 1981) and in Japan (Mogi et al., 1989), suggesting that this type of hydrologic response may be common in hydrothermal areas. Such changes in discharge may also be expected to cause changes in the temperature and the chemical composition of the streams and hot springs (see Chap. 6).

In convergent tectonic regions, large volumes of pore water may be locked in subducted sediments (Townend, 1997). Sealing may be enacted partly by the presence of low-permeability mud, partly by precipitation of minerals in fractures and pores, and partly by the prevailing compressional stresses in such tectonic settings (Sibson and Rowland, 2003). Earthquakes may rupture the seals and allow pressurized pore water to erupt to the surface and recharge streams. Husen and Kissling (2001) suggest that postseismic changes in the ratio of P-wave and S-wave velocities above

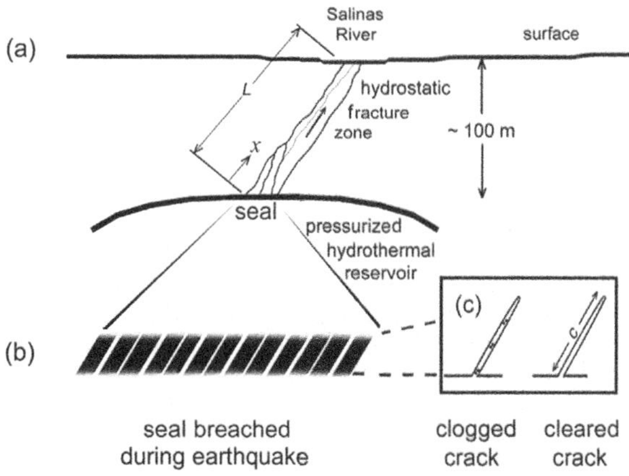

Fig. 4.14 (**a**) Cartoon of the proposed model to explain the hour long increase in streamflow following the 2003 *M* 6.5 San Simeon earthquake, California. Rupturing of the seal of hydrothermal reservoir leads to expulsion of fluid into fracture zone. (**b**) Enlarged diagram of seal with cracks. (**c**) Clogged crack and cleared crack; clearing of a clogged crack significantly increases its permeability and effective length

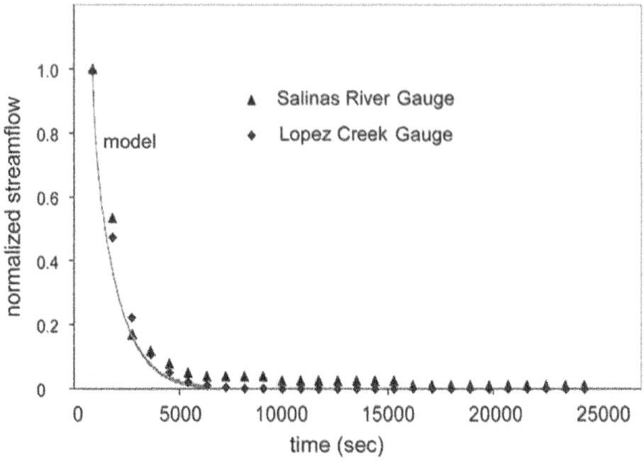

Fig. 4.15 Model prediction (*curve*) compared with normalized stream-gauge data (*triangles* and *diamonds*) for the streamflow changes (From Wang et al., 2004b)

the subducting Nazca Plate reflect fluid migration into the overlying plate following the rupture of permeability barriers. This process may explain the time variations in submarine fluid discharge at convergent margins (Carson and Screaton, 1998). Episodes of high discharge are correlated with seismic activity having features similar to tremor and are not correlated with large regional earthquakes (Brown et al., 2005).

4.7 Concluding Remarks

The different hypotheses listed in this section imply different crustal processes and different water–rock interactions during an earthquake cycle. In most instances, the hydrological models are underconstrained. A reasonable approach is to test the different hypotheses against cases such as that during the Chi-Chi earthquake in which abundant and accurate data are available. We note that a single explanation need not apply for all cases of increased streamflow, so that identifying when and where different mechanisms dominate is important.

References

Angelier, J., H.T. Chu, J.C. Lee, J.C. Hu, F. Mouthereau, C.Y. Lu, B. Deffontaines, S. Lallemand, Y.B. Tsai, J.D. Chow, and D. Bureau, 2000, Geologic knowledge and seismic risk mitigation: Insight from the Chi-Chi earthquakeChi-Chi earthquake (1999): Taiwan. In: C.-H. Lo, and W.-I. Liao (eds.), *Proceedings of International Workshop on Annual Commemoration of Chi-Chi Earthquake, Science Aspect*, pp. 13–24.

Briggs, R.O., 1991, Effects of Loma Prieta earthquake on surface waters in Waddell Valley, *Water Resour. Bull., 27*, 991–999.

Brodsky, E.E., E. Roeloffs, D. Woodcock, I. Gall, and M. Manga, 2003, A mechanism for sustained groundwater pressure changes induced by distant earthquakes, *J. Geophys. Res., 108*(B8), 2390, doi:10.1029/2002JB002321.

Brown K.M., M.D. Tryon, H.R. DeShon, S. Schwartz, and L.M. Dorman, 2005, Correlated transient fluid pulsing and seismic tremor in the Costa Rica subduction zone, *Earth Planet. Sci. Lett., 238*, 189–203.

Carson, B., and E.J. Screaton, 1998, Fluid flow in accretionary prisms: Evidence for focused, time-variable discharge, *Rev. Geophys., 36*, 329–352.

Charmoille A., O. Fabbri, J. Mudry, Y. Guglielmi, and C. Bertrand, 2005, Post-seismic permeability change in a shallow fractured aquifer following a M-L 5.1 earthquake (Fourbanne karst aquifer, Jura outermost thrust unit, eastern France), *Geophys. Res. Lett., 32*, L18406, doi:10.1029/2005GL023859.

Chia, Y.P., Y.S. Wang, H.P. Wu, J.J. Chiu, and C.W. Liu, 2001, Changes of groundwater level due to the 1999 Chi-Chi earthquake (1999) in the Choshui alluvial fan River fan in Taiwan, *Bull. Seism. Soc. Am., 91*, 1062–1068.

Elkhoury, J.E., E.E. Brodsky, and D.C. Agnew, 2006, Seismic waves increase permeability, *Nature, 411*, 1135–1138.

Husen, S., and E. Kissling, 2001, Postseismic fluid flow after the large subduction earthquake of Antofagasta, Chile, *Geology, 29*, 847–850.

King C.-Y., D. Basler, T.S. Presser, W.C. Evans, L.D. White, and A. Minissale, 1994, In search of earthquake-related hydrologic and chemical-changes along hayward fault, *Appl. Geochem., 9*, 83–91.

Lee, C.T., K.I. Kelson, and K.H. Kang, 2000, Hanging wall deformation and its effect on buildings and structures as learned from the Chelungpu faulting in the 1999 Chi-Chi Taiwan earthquake. In: C.-H. Lo, and W.-I. Liao (eds.), *Proceedings of International Workshop on Annual Commemoration of Chi-Chi Earthquake*, pp. 93–104, Science Aspect.

Lee, J.-C., H.-T. Chu, J. Angelier, Y.-C. Chan, J.-C. Hu, C.-Y. Lu, and R.-J. Rau, 2002b, Geometry and structure of northern rupture surface ruptures of the 1999 $M_W = 7.6$ Chi-Chi Taiwan earthquake: Influence from inherited fold belt structures, *J. Struct. Geol., 24*, 173–192.

Leonardi, V., F. Arthaud, A. Tovmassian, and K. Karakhanian, 1990, Tectonic and seismic conditions for changes in spring discharge along the Garni right lateral strike slip fault (Armenian Upland), *Geodin. Acta, 11*, 85–103.

Lin, W.Y., 2000, *Unstable Groundwater Supply after the Big Earthquake*, Taiwan: United Daily (in Chinese).

Manga, M., 2001, Origin of postseismic streamflow changes inferred from baseflow recession and magnitude-distance relation, *Geophys. Res. Lett., 28*, 2133–2136.

Manga, M., E.E. Brodsky, and M. Boone, 2003, Response of streamflow to multiple earthquakes and implications for the origin of postseismic discharge changes, *Geophys. Res. Lett., 30*(5), 1214, doi:10.1029/2002GL016618.

Manga, M., and J.C. Rowland, 2009, Response of alum rock springs to the October 30, 2007 earthquake and implication for increased discharge after earthquakes, *Geofluids, 9*, 237–250.

Manga, M., and C.-Y. Wang, 2007, Earthquake hydrology. In: H. Kanamori (ed.), *Treatise on Geophysics, 4*, Elsevier, Ch. In: *Treatise on Geophysics*, G. Schubert editor, Vol. 4, pp. 293–320.

Mogi, K., H. Mochizuki, and Y. Kurokawa, 1989, Temperature changes in an artesian spring at Usami in the Izu Peninsula (Japan) and their relation to earthquakes, *Tectonophysics, 159*, 95–108.

Montgomery, D.R., and M. Manga, 2003, Streamflow and water well responses to earthquakes, *Science, 300*, 2047–2049.

Montgomery, D.R., H.M. Greenberg, D.T. Smith, 2003, Streamflow response to the Nisqually earthquake, *Earth Planet. Sci. Lett., 209*, 19–28.

Muir-Wood, R., and G.C.P. King, 1993, Hydrological signatures of earthquake strain, *J. Geophys. Res., 98*, 22035–22068.

Roeloffs, E.A., 1998, Persistent water level changes in a well near Parkfield, California, due to local and distant earthquakes, *J. Geophys. Res., 103*, 869–889.

Rojstaczer, S., and S. Wolf, 1992, Permeability changes associated with large earthquakes: An example from Loma Prieta, California, 10/17/89 earthquake, *Geology, 20*, 211–214.

Rojstaczer, S., S. Wolf, and R. Michel, 1995, Permeability enhancement in the shallow crust as a cause of earthquake-induced hydrological changes, *Nature, 373*, 237–239.

Sato, T., R. Sakai, K. Furuya, and T. Kodama, 2000, Coseismic spring flow changes associated with the 1995 Kobe earthquake, *Geophys. Res. Lett., 27*, 1219–1222.

Sibson, R.H., and J.V. Rowland (2003) Stress, fluid pressure and structural permeability in seismogenic crust, North Island, New Zealand, *Geophys. J. Int., 154*, 584–594.

Sorey, M.L., and M.D. Clark, 1981, Changes in the discharge characteristics of thermal springs and fumaroles in the Long Valley caldera, California, resulting from earthquakes on May 25–27, 1980, *U.S. Geological Survey Open-File Report*, 81–203.

Tokunaga, T., 1999, Modeling of earthquake induced hydrological changes and possible permeability enhancement due to the 17 January 1995 Kobe earthquake, Japan, *J. Hydrol., 223*, 221–229.

Townend, J., 1997, Subducting a sponge; minimum estimates of the fluid budget of the Hikurangi Margin accretionary prism, *Geol. Soc. N.Z. Newsletter, 112*, 14–16.

Wang, C.-Y., L.-H. Cheng, C.-V. Chin, and S.-B. Yu, 2001, Coseismic hydrologic response of an alluvial fan to the 1999 Chi-Chi earthquake (1999), Taiwan, *Geology, 29*, 831–834.

Wang, C.-Y., C.H. Wang, and M. Manga, 2004a, Coseismic release of water from mountains: Evidence from the 1999 ($M_W = 7.5$) Chi-Chi, Taiwan, earthquake, *Geology, 32*, 769–772.

Wang, C.-Y., M. Manga, D. Dreger, and A. Wong, 2004b, Streamflow increase due to rupturing of hydrothermal reservoirs: Evidence from the 2003 San Simeon earthquake (2003), California, earthquake, *Geophys. Res. Lett., 31*, L10502, doi:10.1029/2004GL020124.

Wang C.-Y., M. Manga, and A. Wong, 2005, Floods on Mars released from groundwater by impact, *Icarus, 175*, 551–555.

Wang, C.-Y., 2007, Liquefaction beyond the near field, *Seismo. Res. Lett., 78*, 512–517.

Wakita, H., 1975, Water wells as possible indicators of tectonic strain. *Science, 189*, 553–555.

Water Resource Bureau, 2000, *Report and analysis of changes in surface and ground water due to the 9/21 Chi-Chi Earthquake*: Taipei, Water Resource Bureau, Ministry of Economic Affairs, Taiwan (in Chinese), 37 p., with 4 appendices.

Yan, H.R., 2001, Water problems in the Yuanlin Mountains: Changes in groundwater after the Chi-Chi earthquake (1999), Taiwan, Chinese Daily, (in Chinese).

Chapter 5
Groundwater Level Change

Contents

5.1 Introduction

Water-level changes in wells during and after earthquakes have been reported since the time of antiquity (e.g., Institute of Geophysics – CAS, 1976) and are the most widely documented changes among all earthquake-induced hydrologic phenomena. Since the early twentieth century, instrumental records of water-level changes during earthquakes have become available and significant advances in quantitative analysis have been made (Cooper et al., 1965; Liu et al., 1989; Roeloffs, 1998; King et al., 1999; Roeloffs et al., 2003; Brodsky et al., 2003; Matsumoto et al., 2003; Montgomery and Manga, 2003; Wang et al., 2003; Sato et al., 2004; Kitagawa et al., 2006; Sil and Freymueller, 2006; Elkhoury et al., 2006; Wang and Chia, 2008; Wang et al., 2009). Results of these analyses have significantly contributed to our understanding of the responses of hydrogeologic system to earthquakes.

C.-Y. Wang, M. Manga, *Earthquakes and Water*, Lecture Notes in Earth Sciences 114, DOI 10.1007/978-3-642-00810-8_5, © Springer-Verlag Berlin Heidelberg 2010

In general, two approaches have been followed in documenting earthquake-induced groundwater level changes: One relies on data from a single well that responds to many earthquakes (e.g., Roeloffs, 1998; Matsumoto et al., 2003; Brodsky et al., 2003); the other examines data from many wells that respond to a single earthquake (Chia et al., 2001, 2008; Wang et al., 2001, 2004; Chia et al., 2008; Wang and Chia, 2008). Using the records from a single well in central California, Roeloffs (1998) identified three categories of groundwater level responses: In the near field, groundwater level shows step-like increases (Fig. 5.1b). In the intermediate field, groundwater level changes are more gradual and can persist for hours to weeks (Fig. 5.1c). At even greater distances (the far field), only transient oscillations of the water-level were documented (Fig. 5.10).

A network of densely distributed monitoring wells in central Taiwan near the epicenter revealed the occurrence of several distinct types of coseismic and postseismic responses after the 1999 Chi-Chi earthquake. In the immediate vicinity (within a few km) of the ruptured fault, groundwater level showed step-like decreases (Fig. 5.1a) followed by either an exponential increase or decrease with time. Further away, but still in the near field, groundwater level showed step-like increases followed by an exponential decrease with time (Fig. 5.1b). Still further away, groundwater level showed sustained changes. Thus there is a general agreement between the observations from the two approaches. The absence of the step-like decrease in Roeloffs (1998) may be due to absence of faults that ruptured in the immediate vicinity of the observation well. The absence of groundwater oscillations in Wang et al. (2004) may partly be due to the fact that most wells were in the near field of the Chi-Chi earthquake and partly because the sampling rate of the recording instruments was too low (once per hour) to register groundwater oscillations. The fact that the two different approaches, applied in different parts of the world, have led to consistent results provides some reassurance that conclusions may be generally valid.

The major advantages of studying the response of a single well to multiple earthquakes is that the well itself is often carefully calibrated, thus the various extraneous effects on the groundwater level records can often be eliminated and very small changes in the groundwater level source may be detected at great distances from the earthquake (e.g., Roeloffs, 1998; Roeloffs et al., 2003; Brodsky et al., 2003). Furthermore, since the geology of the well site does not change during different earthquakes, the complication introduced from the geology of the well site may often be eliminated and the groundwater records can often be used to effectively discriminate between different models of the causal mechanisms of groundwater level changes. There are two major advantages of using a dense network of monitoring wells (e.g., Chia et al., 2001; Wang et al., 2001, 2004; Chia et al., 2008). First, if the network of wells is both densely and broadly distributed, a continuous spatial distribution of groundwater-level response to an earthquake may be constructed. Second, if a dense seismic network is also in existence, the spatial relationship between seismic waves and groundwater level changes may be examined.

In the following we discuss separately the groundwater level changes in the near field, intermediate field, and far field of earthquakes, each illustrated with examples. We follow these with a discussion of the causal mechanisms. In addition we

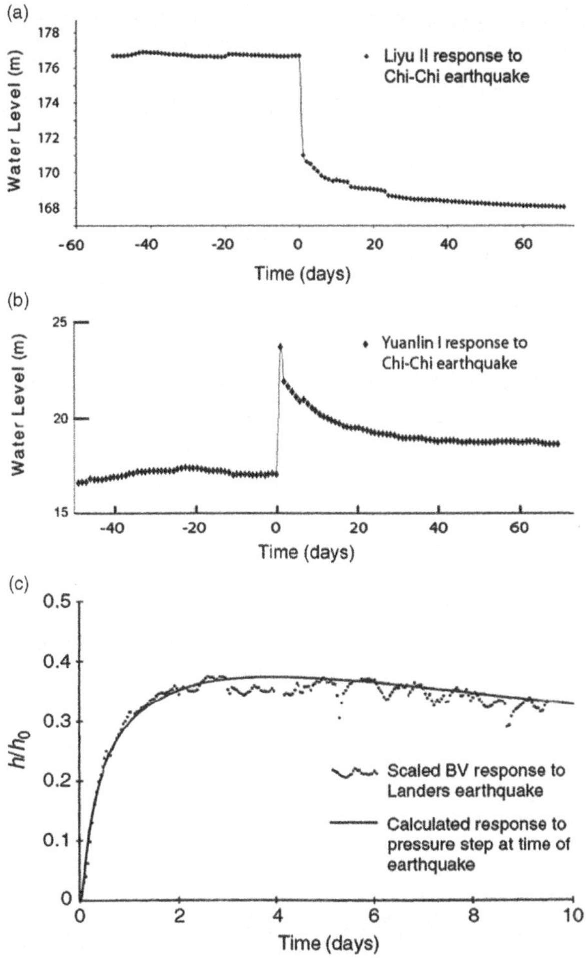

Fig. 5.1 (**a**) Negative water-level change in the Liyu II well during the 1999 M_W7.5 Chi-Chi earthquake (occurring at $t = 0$) in Taiwan. The well is ~20 km from the hypocenter but is only ~5 km from the surface rupture of the causative fault. The step-like coseismic water-level decrease is –5.94 m. (**b**) Positive water-level change in the Yuanlin I well during the 1999 M_W7.5 Chi-Chi earthquake (occurring at $t = 0$). The well is ~25 km from the hypocenter, thus in the near field, and ~13 km from the surface rupture of the causative fault. The step-like coseismic water-level increase is +6.55 m. (**c**) Scaled groundwater level response in a well in central California to the 1992 M 7.3 Landers earthquake 433 km from the well. The observed water-level changes, shown with data points, can be modeled by a coseismic, localized pore pressure change at some distance from the well, shown by the solid curve ((**a**) and (**b**) from Wang and Chia, 2008, (**c**) from Roeloffs, 1998)

discuss the recent discovery of groundwater oscillations in response to S waves and Love waves, and of pore pressure changes on the seafloor. We end the chapter with an analysis of the postseismic recession of the groundwater level and compare the results from the analysis of the postseismic baseflow recession in the previous chapter.

5.2 Step-like Changes in the Near Field

5.2.1 Observations

In the near field, i.e., the area around the hypocenter within a distance of ~1 ruptured fault length, groundwater level often shows step-like changes during earthquakes (Wakita, 1975; Quilty and Roeloffs, 1997; Chia et al., 2001; Wang et al., 2001) and changes in excess of 10 m in amplitude are not uncommon. In unconsolidated sedimentary aquifers, most documented changes are positive (Fig. 5.1b). In the immediate vicinity (within a few km) of the ruptured fault, however, negative changes were reported (Fig. 5.1a). In some cases, wells installed in fractured igneous aquifers show a characteristic pattern of positive and negative changes of groundwater level similar to that of the coseismic static strain change (Fig. 5.2). In other cases, however, such a pattern is contradicted by observation (e.g., Wang et al., 2001; Koizumi et al. 2004).

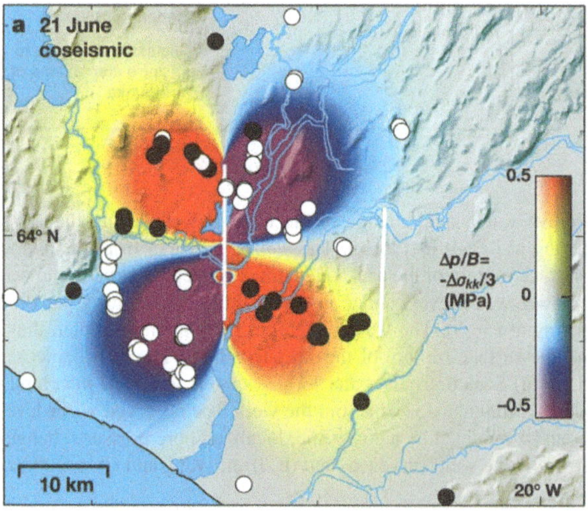

Fig. 5.2 Calculated pore-pressure change based on the coseismic volumetric strain and theory of poroelasticity (*color map*) and observed coseismic water-level changes (*circles*) following a 2000 M 6.5 strike-slip earthquake in Iceland. *Black* and *white circles* indicate water-level increases and decreases, respectively. The *white line* shows the mapped surface rupture (From Jonsson et al., 2003)

As noted in Chap. 2, a dense network of monitoring wells was installed on an alluvial fan (the Choshui River fan) near the Chi-Chi epicenter (Fig. 2.12) prior to the earthquake. About 200 wells captured the groundwater level changes during and after the earthquake. In addition, several rain gauges installed around the fan provide continuous records of precipitation in the area. The close proximity to a large earthquake and the dense network of hydrological stations made this dataset one of the most comprehensive and systematic to study the spatial distribution of the hydrologic response in the near field. For the sake of better interpreting the observations, a brief summary of the subsurface hydrogeology of the Choshui River alluvial fan is needed. Figure 5.3 shows a simplified hydrogeological cross-section along the line AB in Fig. 4.3. It shows that the alluvial fan consists of subhorizontal layers of unconsolidated Holocene and Pleistocene sediments. Three distinct aquifers may be distinguished (Fig. 5.3): Aquifer I, the topmost aquifer, is partly confined and partly unconfined, while aquifers II and III are confined. To the east of the alluvial fan is a fold-and-thrust belt (the Western Foothills) that consists of consolidated Pleistocence sedimentary rocks which were pervasively faulted and fractured by past earthquakes.

Water level in most wells is recorded by digital piezometers and the data are logged at 1-h intervals. Some wells are equipped with data logger operating at

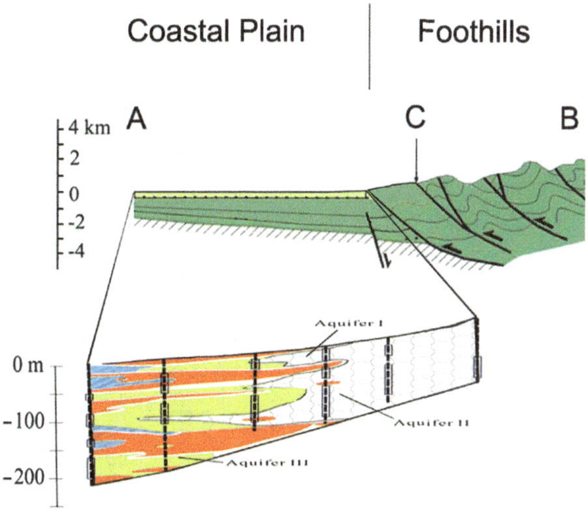

Fig. 5.3 Simplified cross-section along the line AB in Fig. 4.3, showing the general relationship between the Coastal Plain and the fold-and-thrust Foothills. The surface trace of the ruptured Chelungpu fault is marked by the letter C. The enlarged inset shows three aquifers in the Choshui River alluvial fan. Massive gravel beds occur in proximal area; away from proximal area, gravel beds decrease in thickness, while coarse sands and fine sands increase in proportion and interfinger with gravel beds; further away, silty sands and silty clays increase in proportion and eventually dominate the distal margin of fan. Three aquifers can be identified from top (Aquifer I) to *bottom* (Aquifer III). The *vertical dashed lines* represent boreholes. *Vertical axis* on *left* gives elevation above (and *below*) mean sea-level in m (Reproduced from Wang et al., 2005)

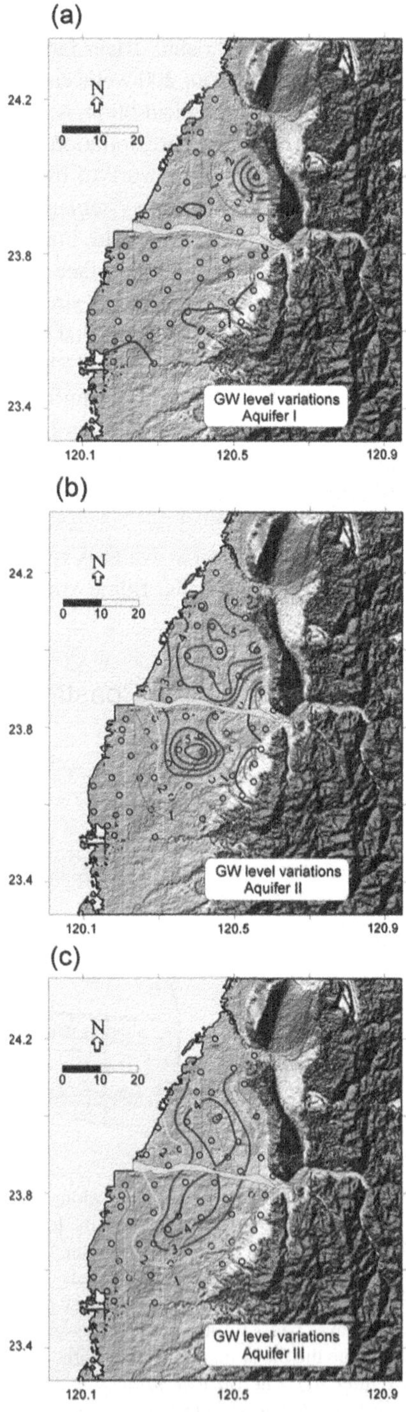

Fig. 5.4 Distribution of the coseismic changes in groundwater level in the Choshui River fan during the Chi-Chi earthquake. (**a**) Contours (m) of groundwater level change in the uppermost aquifer (Aquifer I). (**b**) Contours (m) of groundwater level changes in a confined aquifer (Aquifer II). (**c**) Contours (m) of groundwater level changes in a lower confined aquifer (Aquifer III). Groundwater monitoring stations are shown in *open circles*, epicenter of Chi-Chi earthquake in *star*, and the ruptured fault in discontinuous *red* traces (From Wang et al., 2005)

high rate (up to 1 Hz), while others are equipped with analogue data loggers which provide continuous records. Piezometer readings in the well are converted to groundwater level with an accuracy of 1 cm. The resolution of the reading, on the other hand, is finer by an order of magnitude.

Figure 5.4 shows the coseismic changes in groundwater level in the three aquifers in the Choshui River fan during the Chi-Chi earthquake. There are important differences in the coseismic responses among different aquifers. In the uppermost, partially confined aquifer, the coseismic changes are generally small except in an area on the northeastern edge of the alluvial fan where positive changes occurred (Fig. 5.4a). This area of positive water-level change is closely associated with the occurrence of liquefaction on the fan during the Chi-Chi earthquake (see Fig. 2.13 and Sect. 2.5.1).

The distribution of coseismic changes in water level in the two confined aquifers (Fig. 5.4b, c), on the other hand, showed an entirely different pattern of coseismic groundwater level changes and there was no association of these changes with the occurrence of liquefaction on the surface. Instead, the water-level rise in these aquifers showed an unexpected pattern of increase with distance away from the ruptured fault, reaching a maximum at distances of 20–30 km from the fault, and then decreased at greater distances (Chia et al., 2001; Wang et al., 2001).

5.2.2 *Causal Mechanisms*

5.2.2.1 Static Strain Hypothesis

Both the sign and magnitude of the coseismic water-level changes can be compared with those predicted from models to provide insight into the causal mechanisms. In some cases the coseismic groundwater level changes can be explained by the coseismic static strain and pore pressure change predicted by poroelastic theory, see Appendix D (e.g., Wakita, 1975; Roeloffs, 1996; Ge and Stover, 2000; Jonsson et al., 2003). In such cases, the water level will rise in zones of contraction, and fall in regions of dilation. Figure 5.2, from Jonsson et al. (2003), shows a pattern of groundwater level changes that mimics the pattern of coseismic volumetric strain after a strike-slip event in Iceland. An analogous correlation of volumetric strain and water-level changes was found after the 2003 M 8.0 Tokachi-oki thrust event in Japan (Akita and Matsumoto, 2004). Similar inferences were made by others (e.g., Wakita, 1975; Igarashi and Wakita, 1991; Quilty and Roeloffs, 1997). This pattern of water-level change, however, was contradicted in other instances where coseismic groundwater level changes were documented by densely distributed monitoring wells. Following the 1999 M_W7.6 Chi-Chi earthquake in Taiwan, for example, the water level rose in most wells where the coseismic change of volumetric strain is positive (dilatation), and fell in other wells where the volumetric strain change is negative (contraction) (Fig. 5.5, from Koizumi et al., 2004); thus the observed responses is just the opposite from the pattern predicted from the hypothesis of coseismic static volumetric strain.

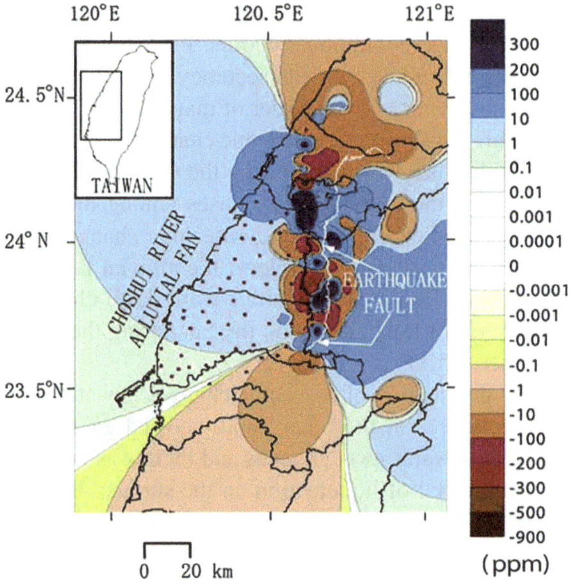

Fig. 5.5 Distribution of coseismic volumetric strain changes calculated from a fault model for the Chi-Chi earthquake. Positive and negative values indicate dilatation and contraction, respectively. *Black dots* are the locations of observation wells (From Koizumi et al., 2004)

5.2.2.2 Undrained Consolidation Hypothesis

A different model for groundwater level change during earthquakes is that ground shaking causes sediments around a well to consolidate or dilate, leading to step-like changes in pore pressure and changes in groundwater level in the well (Wang et al., 2001; Wang and Chia, 2008). During the last half century, earthquake engineers have performed numerous laboratory experiments to study consolidation of saturated sediments under cyclic shearing (for summaries, see National Research Council, 1985; Ishihara, 1996). The results of these experiments show that, when sediments are subjected to cyclic shearing, they begin to consolidate if the shear strain magnitude exceeds $\sim 10^{-4}$ (Dobry et al., 1982; Vucetic, 1994). If the deformation is undrained, as expected during seismic shaking, pore pressure increases when the shear strain exceeds about 10^{-4}. However, when the amplitude of shaking becomes so large to exceed some critical threshold, cracks and pores may dilate and new fractures may form, leading to increased porosity and decreased pore pressure (Fig. D.2, from Luong, 1980). A critical state with no volumetric change occurs between these two states. These laboratory results are qualitatively consistent with the field observation of the water-level changes on the Choshui River fan during the Chi-Chi earthquake, i.e., immediately adjacent to the ruptured fault where the seismic shaking is the strongest, the water level shows a coseismic decline (Fig. 5.1a); within the near field yet at some distance from the fault, the water level shows a coseismic rise (Fig. 5.1b). Wang et al. (2001) used the critical state concept to explain the increase in the groundwater level change with distance from the ruptured fault (Fig. 5.4b, c).

5.2.2.3 Energy to Initiate Undrained Consolidation

A great number of laboratory experiments have been carried out by geotechnical engineers to study the consolidation of sediments (e.g., Seed and Lee, 1966; Dobry et al., 1982; Vucetic, 1994; Hsu and Vucetic, 2004). Based on experimental data for a wide variety of saturated sediments and a wide range of confining pressures, Dobry et al. (1982), Vucetic (1994) and Hsu and Vucetic (2004) showed that pore pressure begins to increase when sediments are sheared at a strain amplitude of 10^{-4} for 10 cycles. Yoshimi and Oh-Oka (1975) further showed that this result may not be significantly affected by the frequency of loading. Here we show that the laboratory results may be used to calculate the amount of dissipated energy required to cause undrained consolidation; the magnitude of this energy may be compared with the seismic wave energy in the field at different distances from the earthquake source.

The dissipated energy density required to initiate undrained consolidation in saturated sediments may be estimated from the experimental time histories of shear stress τ and strain γ by performing the following integration:

$$e_d = \int_0^t \tau \, d\gamma \tag{5.1}$$

where the integration extends from the beginning of the cyclic loading to the onset of pore-pressure increase. Although this equation is the same as Eq. 2.1 (Sect. 2.3.2) used to calculate the dissipated energy required to initiate liquefaction, the integration procedures in the two situations are entirely different. In the present case, both the shear strain and the dissipation are small and the stress and strain relation is nearly linear. Thus we may express the cyclic experimental stress and strain in the form $\tau = \tau_o \sin\theta$ and $\gamma = \gamma_o \, sin \, (\theta + \varphi)$, respectively, where τ_o and γ_o are the corresponding amplitudes and φ is the phase angle between stress and strain. Integrating Eq. (5.1) we obtain, for small φ (Table 5.1),

$$e_d = \frac{N\pi}{2} \tau_o \gamma_o \sin \varphi \cong \frac{N\pi}{2} \mu \gamma_o^2 \varphi \tag{5.2}$$

where $N \sim 10$ is the usual number of cycles adopted in experimental studies, μ is the shear modulus measured at strain amplitude of 10^{-4} and φ is equal to the damping ratio (Ishihara, 1996). We list in Table 5.1 the available experimental data for the shear modulus and the damping ratio, together with the calculated energy density e_d values. The calculated e_d range from ~0.1 to ~5 J/m^3. The range of dissipated energy density required to initiate undrained consolidation reflects the different sensitivity for sediments to consolidate under cyclic loading and the different confining pressures under which the measurements were made (e.g., Ishihara, 1996). Undrained dilatation would require a greater amount of energy to initiate (Luong, 1980); however, the experimental data is not sufficient to evaluate the threshold energy quantitatively.

5.2.2.4 Seismic Energy Density and Groundwater-Level Change

In Chap. 2 (Sect. 2.4.1) we defined the seismic energy density $e(r)$ as the maximum seismic energy available to do work in a unit volume at the hypocentral distance r and derived an empirical relationship among e, r and the earthquake magnitude M:

$$\log r = 0.48 M - 0.33 \log e(r) - 1.4 \tag{2.7}$$

Table 5.1 Shear moduli and damping ratios of sediments, determined under laboratory conditions and cyclically sheared to a strain amplitude of 10^{-4}. Dissipated energy is calculated using Eq. (5.2) with $N = 10$

Samples	Shear modulus* (MPa)	Damping ratio*	Dissipated energy (J/m^3)
Undisturbed Fujisawa sand	100	0.07	1
Disturbed Fujisawa sand	40	0.02	0.1
Undisturbed Tokyo gravel, at a confining pressure of 300 Kpa	300	0.04	2
Undisturbed Tokyo gravel, at a confining pressure of 500 Kpa	600	0.05	5
Reconstituted Tokyo gravel, at a confining pressure of 300 Kpa	200	0.07	2
Reconstituted Tokyo gravel, at a confining pressure of 500 Kpa	400	0.08	5

*Experimental values for the shear modulus and the damping ratio were read from figures on pages 143 and 145 of Ishihara (1996), and are thus approximate.

Here we plot contours of constant e on a diagram of log r versus M (Fig. 5.6) together with the documented coseismic groundwater-level changes. The figure shows that the documented groundwater-level change occurs across seven orders of magnitude in seismic energy density.

Hazirbaba and Rathje (2004) showed that the threshold strain required to initiate undrained consolidation in the laboratory is the same as that in the field. Thus it may be justified to compare the laboratory-based dissipated energy required to initiate undrained consolidation with the seismic energy density in the field. In Fig. 5.6, we mark the range of laboratory-based dissipated energy required to initiate undrained consolidation by a hatched band. Below the hatched band, the seismic energy density is greater than the threshold energy required to initiate undrained consolidation; above the hatched band, the seismic energy density is too small to initiate undrained consolidation; within the hatched band, the seismic energy may be enough to cause consolidation in the more sensitive sediments, but not in the less sensitive ones.

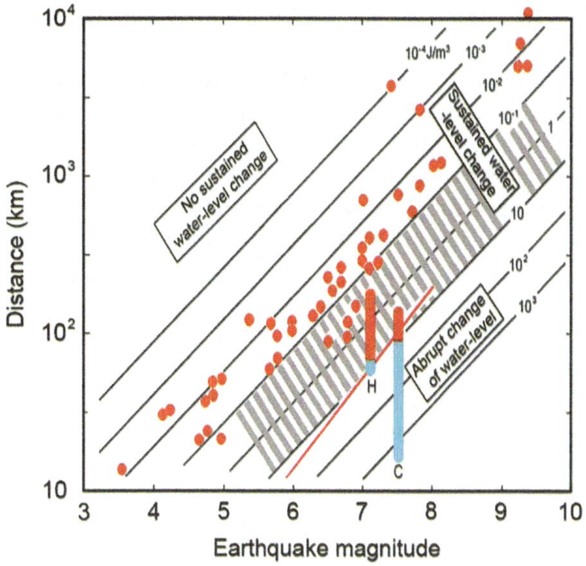

Fig. 5.6 Hypocentral distance (in km) of groundwater-level changes plotted against earthquake magnitude. It shows contours of constant seismic energy density and the domains where different types of coseismic water-level responses occur. The hatched band shows the laboratory-determined dissipated energy required to initiate undrained consolidation. Above the hatched band in the intermediate field, the seismic energy density may be less than the laboratory-determined threshold energy required to initiate undrained consolidation; below the hatched band in the near field, the seismic energy density may be large enough to initiate undrained consolidation; within the hatched band, the seismic energy may be sufficient to initiate consolidation in the more sensitive sediments, but not in the less sensitive ones. The *red line* shows the hypocentral distance equivalent to one ruptured fault length as a function of earthquake magnitude (Wells and Coppersmith, 1994). Sustained water-level changes in a global dataset (see text) are shown in *red circles*; some represent multiple wells located in close proximity. Colored bars show the range of hypocentral distances for water-level changes during the 1999 Chi-Chi M7.5 earthquake (labeled C) and during the 2006 M7.0 Hengchun earthquake (labeled H); red shows distances where both positive and negative changes occurred; *blue* shows distances where most changes were step-like and positive (From Wang and Chia, 2008)

5.3 Sustained Changes in the Intermediate Field

5.3.1 Observations

In the intermediate field, defined here as the area more than 1 ruptured fault length away from the earthquake source but within a radius of 10 ruptured fault lengths, groundwater-level changes are usually characterized by a more gradual onset and can last for hours to weeks (e.g., Fig. 5.1c). Thus some sustained changes in groundwater level may appear 'step like' if the time scale is compressed or if the sampling rate is low.

From the BV well data, Roeloffs (1998) showed that the epicentral distance for earthquakes that induced sustained groundwater level changes is bounded by an empirical relation: $M = -3.91 + 1.82 \log R$, where M is earthquake magnitude and R the threshold distance is in meters. Similarly, King et al. (1999) obtained a relation $M = -7.5 + 2.5 \log R$ for the sustained groundwater level changes documented by a cluster of wells in Japan, and Matsumoto et al. (2003) found $M = -6.90 + 2.45 \log R$ for the data documented by the Haibara well in Japan. The differences between these relations may reflect the small number of data used in the different analyses and/or the different geology of the individual well sites where the data were documented, or both.

In order to increase the number of data, we collected data from published sources up to 2006 (Roeloffs, 1998; King et al., 1999; Roeloffs et al., 2003; Brodsky et al., 2003; Matsumoto et al., 2003; Sato et al., 2004; Kitagawa et al., 2006; Sil and Freymueller, 2006) together with some unpublished data in a single list (Appendix E.2), referred to here as the 'global dataset', and plot them as red circles in Fig. 5.6. Most of the wells were installed in Quaternary or Tertiary sediments and sedimentary rocks. A few were installed in fractured igneous rocks. The difference in geology may have caused some scatter in the data. For example, the data points for M7.4 at 3811 km and M7.9 at 2772 km were from a well constructed in fractured granite (Brodsky et al., 2003); the high sensitivity of this well may explain the great distance at which the groundwater responded to earthquakes. Most of these data, plotted as red circles in Fig. 5.6, fall above the hatched band where the seismic energy density is below the threshold required to initiate undrained consolidation. Thus a new mechanism, other than undrained consolidation, is required to explain the documented water-level changes at such distances (Fig. 5.6).

5.3.2 *Causal Mechanisms*

5.3.2.1 Proposed Hypotheses

The hypothesis of static poroelastic volumetric strain induced by earthquakes, initially proposed by Wakita (1975) to interpret groundwater-level changes in the near field, was also used to interpret changes in the intermediate field (e.g., Chia et al., 2008). Undrained consolidation, initially proposed to explain groundwater-level changes in the near field (Wang et al., 2001; Sect. 5.2.2), was also used to explain pore-pressure changes beyond the near field (e.g., Holzer and Youd, 2007). Several authors suggested that seismic waves may enhance rock permeability by removing precipitates and colloidal particles from clogged fractures, which in turn may lead to redistribution of pore pressure and changes of water level in areas near a local pressure source (Matsumoto et al., 2003; Brodsky et al., 2003; Wang and Chia, 2008). The local pressure source could be the groundwater head in elevated terrains

(Rojstaczer et al., 1995) or produced by localized liquefaction (Roeloffs, 1998). In the following section we show some field tests of these hypotheses.

5.3.2.2 Field Tests of Hypotheses

Roeloffs (1998) suggested that a viable interpretation for the sustained groundwater-level changes must account for the following three observations: (1) the groundwater level response is sustained after the seismic vibration has stopped, (2) the groundwater level response is relatively large, even for distant earthquakes, and (3) the groundwater level always rises in some wells but falls in other wells, regardless of the locations or focal mechanisms of the earthquakes. According to the model of enhanced permeability, the chance of a positive or a negative change in the water level in a well should be the same since the source (or sink) could occur either up-gradient or down-gradient of the well. Thus, if a sufficiently large number of observations are available, the model of enhanced permeability would predict a statistically random occurrence in the sign of the water-level changes. On the other hand, the model of undrained consolidation of sediments around a well would predict mostly a positive change in water level, unless the well is so close to the ruptured fault that the ground experiences coseismic dilation. The strong contrast between the predicted groundwater behaviors permits a test of the two mechanisms by using the documented field observations.

Using the records from a single well in central California, Roeloffs (1998) found that some sustained groundwater level changes showed the opposite sign to that expected from the static poroelastic volumetric strain induced by the earthquake. Another example inconsistent with the static volumetric strain hypothesis occurred in Japan, following the 2004 M 9 Sumatra earthquake more than 5000 km away, where Kitagawa et al. (2006) found that only half of the monitoring wells equipped with strain instruments recorded water-level changes consistent with the measured coseismic strain. Finally, the magnitude of water-level changes at such large distances are often much greater than those predicted from the static strains by the poroelastic theory (e.g., Igarashi and Wakita, 1995; Itaba and Koizumi, 2007; Manga and Wang, 2007).

Figure 5.7a shows, for the global dataset, the sign and amplitude of the sustained groundwater level changes as functions of distance to the earthquake source. The plot reveals a random distribution of the magnitude and the signs of these changes. Thus the data is consistent with the enhanced permeability model.

Figure 5.7b shows the data from a dense network of monitoring wells in central Taiwan near the epicenter of the 1999 M7.5 Chi-Chi earthquake (Fig. 5.4). The data provides a nearly continuous change of the signs and amplitudes of water-level changes from the vicinity of the ruptured fault to a distance of 160 km. The groundwater-level change is predominantly positive except in the immediate neighborhood of the ruptured fault, marked by the downward arrow, where the groundwater-level becomes negative.

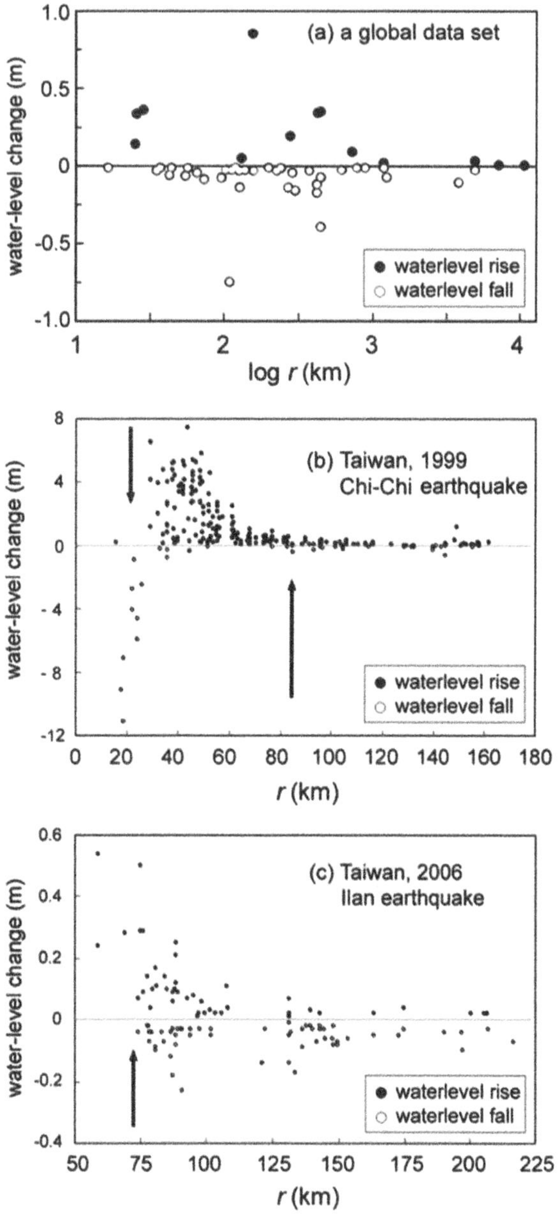

Fig. 5.7 Amplitude and sign of water-level changes during earthquakes plotted as functions of the hypocentral distance. (**a**) Water-level changes in a global dataset of sustained groundwater level changes (see text). The *horizontal line* gives the mean of all the groundwater level changes (−0.004 m) and the standard error is 0.025 m. (**b**) Water-level changes during the 1999 Chi-Chi earthquake. The upward-pointing *arrow* shows the distance equal to one ruptured fault length. The downward-pointing *arrow* shows the location of the ruptured fault. Note that nine wells near the downward-pointing *arrow* documented abrupt decreases of water level as illustrated in Fig. 1b. These wells are all located within 5 km of the surface rupture of the causative fault. (**c**) Water-level changes during the 2006 Hengchun earthquake. The upward-pointing *arrow* shows the distance equal to one ruptured fault length. Note that at distances beyond one ruptured fault length, the sign of water-level changes is randomly distributed (From Wang and Chia, 2008)

Figure 5.7c shows the water-level changes in a large number of monitoring wells in southern Taiwan, documented during the 2006 Hengchun earthquake off the southern coast of Taiwan. Most wells were at distances beyond one ruptured fault length, marked by the upward arrow, and showed both positive and negative changes at further distances (Fig. 5.7c). A few wells at closer distances show only positive changes.

In Fig. 5.6 we show the ranges of hypocentral distances for water-level changes during the Chi-Chi and the Hengchun earthquakes (colored bars marked by C and H, respectively). The change in color in each bar marks the distance of transition from locations with only positive changes to locations with both positive and negative changes (Fig. 5.7b, c). In both cases, this transition occurs close to the lower bound of the hatched band, i.e., at e $\sim$10 J/m^3, consistent with the hypothesis that it corresponds to a change in mechanism from undrained-consolidation to enhanced-permeability.

All in all, the observations are consistent with the hypothesis that, in the near field, undrained consolidation or dilatation is the dominant mechanism for the coseismic changes of groundwater level; in the intermediate field, earthquake-enhanced permeability is the dominant mechanism. These mechanisms provide simple explanations for why the water-level changes are step-like in the near field and more gradual (and sustained) in the intermediate field and why a given well in the intermediate field shows consistently positive or consistently negative water-level changes during different earthquakes (e.g., Roeloffs, 1998; Matsumoto et al., 2003). The occurrence of undrained volumetric changes around a well in the near field causes an immediate change in pore pressure, thus a step-like change in water level. On the other hand, an enhanced permeability connects a well to a nearby pressure source (or sink); the rise time of the water level in the well will depend on the square of the distance between the well and the pressure source or sink; thus the water level changes in such cases are gradual and sustained, as explained by previous authors (Roeloffs, 1998; Brodsky et al., 2003). Furthermore, since the same source (or sink) and permeable passageway may be activated during different earthquakes, one may expect a persistent sign of water-level change during different earthquakes.

We note that enhanced permeability may occur both in the near field and in the intermediate field. In the near field, however, the abrupt changes of groundwater level due to undrained dilatation or consolidation may be so large in amplitude that they obscure the sustained groundwater level changes which are usually of smaller amplitude (<1 m). Thus the sustained changes can be clearly detected only in the intermediate field where the mechanism of undrained volume changes is no longer important. Finally, the enhanced permeability may decrease with time due to re-clogging of the passageways by hydrogeochemical and/or biogeochemical processes.

So far we have discussed the lateral variations in the groundwater level following earthquakes because in most cases no depth variation is measured. However, when individual aquifers at different depths in a given location are monitored by closely spaced wells, the responses of the aquifers to earthquakes may show significant differences (Chia et al., 2008), as may be expected from the different hydro-mechanical properties of the aquifers due to their different ages, overburden and lithology. Sometimes the different responses may be explained by a change in the vertical permeability of the barrier (aquitard) between two adjacent aquifers (Wang and Chia, 2008; see also Fig. 2.7).

5.3.2.3 Earthquake- Enhanced Permeability Experiment

Responding to Earth's solid tides, groundwater in aquifers continually flows in and out of some wells. The phase lag between the solid tides and the observed water level provides a means to monitor the permeability of aquifers. The data in Fig. 5.8 (from Elkhoury et al., 2006) from two wells in southern California over the past 20 years show repeated stepwise changes in phase at the time of earthquakes, marked by the vertical lines, followed by a gradual recovery of the phase to the pre-earthquake values. Note that all earthquakes produce a decrease in the phase lag, implying an increase in aquifer permeability, regardless of the sign of the earthquake-induced

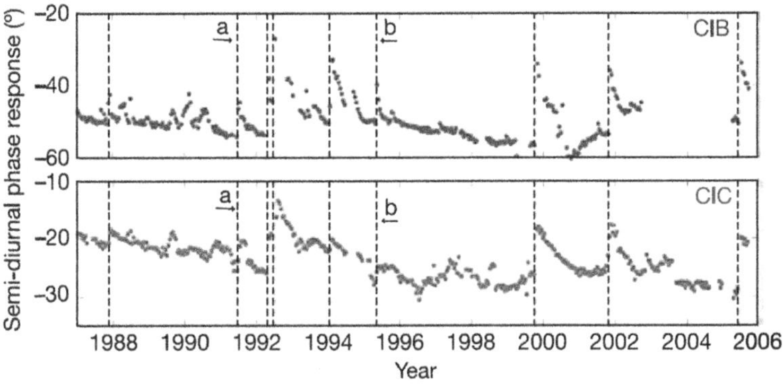

Fig. 5.8 Phase of the semi-diurnal tides for the water levels in two wells in southern California relative to the tidal strain. Transient changes of the phase are clearly evident at the time of earthquakes, as shown by the *vertical lines* (From Elkhoury et al., 2006)

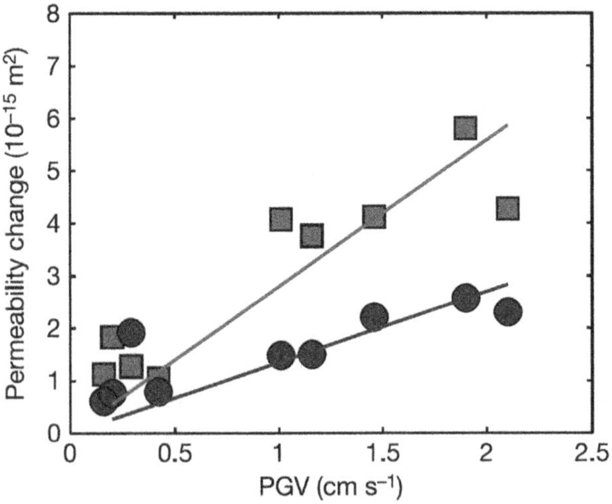

Fig. 5.9 Increased permeability of the aquifers at the two wells plotted against the peak ground velocity (PGV) of ground motion during earthquakes. The maximum increases correspond to a factor of 5–6 increase in aquifer permeability (From Elkhoury et al., 2006)

static strain at the well sites (Elkhoury et al., 2006). The last observation clearly shows that the change in permeability was not caused by the static strain, but by the dynamic strain.

The aquifer permeability may be calculated from the amplitude response and phase lag of groundwater oscillations with respect to the tidal forcing (Hsieh et al., 1987; Appendix B.9). Figure 5.9 shows the increases of permeability during earthquakes evaluated at the two wells as functions of the peak ground velocity. For each aquifer, a linear trend is apparent between the increased permeability and the peak ground velocity. In other words, the amount of increase in permeability appears to be proportional to the intensity of ground shaking.

5.4 Groundwater Oscillations in the Far Field

Groundwater level in wells can amplify ground motions and, when recorded at high enough frequencies, often shows oscillations associated with long-period Rayleigh waves. Such 'hydroseismograms' have been documented since the early days of seismometer use (Blanchard and Byerly, 1935) and groundwater level fluctuations as large as 6 m (peak-to-peak amplitude) were recorded in Florida, thousands of kilometers away from the epicenter, during the 1964 Alaska earthquake (Cooper et al., 1965).

Figure 5.10 (from Brodsky et al., 2003) shows the groundwater level in a well in Grants Pass, Oregon, following the 2002 *M* 7.9 Denali earthquake 3100 km away. Also shown is the vertical component of ground velocity measured on a broadband seismometer located adjacent to the well. Comparing the water-level record

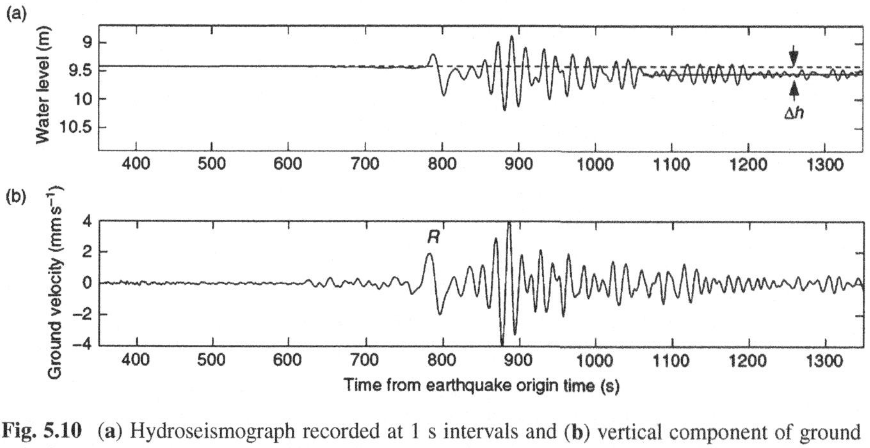

Fig. 5.10 (**a**) Hydroseismograph recorded at 1 s intervals and (**b**) vertical component of ground velocity measured on an broadband seismometer at a well in Grants Pass, OR, following the 2002 M7.9 Denali earthquake located 3100 km away. Δh shows the 12 cm permanent change in water level that followed the passage of the seismic waves; R indicates Rayleigh waves (From Brodsky et al., 2003)

and the seismogram shows a close correlation of the water-level oscillation with Rayleigh waves. Large water-level oscillations such as these at great distances occur when the geometric (water depth in the well, well radius, aquifer thickness) and the hydrogeologic (transmissivity) properties have the right combination (e.g., Cooper et al., 1965; Liu et al., 1989). The commonly accepted mechanism is that seismic waves may cause aquifers to expand and contract which in turn may cause pore-pressure to oscillate. If the aquifer has high enough transmissivity, the pore-pressure oscillations may induce fluid flow into and out of the well, which in turn may set up resonant motions in the water column. Such responses can be determined theoretically by solving the coupled equations for pressure change in the aquifer, groundwater flow into and out of the aquifer, and flow within the well (e.g., Cooper et al., 1965; Liu et al., 1989). In general, high transmissivity favors large amplitudes. Kono and Yanagidini (2006) found that in closed (as opposed to open) borehole wells, pore pressure variations are consistent with an undrained poroelastic response (Appendix D) at seismic frequencies.

Brodsky et al. (2003) further noticed that the groundwater level oscillations in some wells were associated with an increase in the magnification factor, defined as the ratio of the spectrum of groundwater oscillation to the spectrum of the particle velocity of Rayleigh waves. This increase was interpreted to indicate an enhancement of the permeability of a fractured aquifer due to the removal of loose particles from fractures by the oscillating flow.

5.5 Role of S waves and Love Waves on Groundwater Oscillations

In the previous section we showed that groundwater oscillations may occur in response to the volumetric strain associated with Rayleigh waves. After the 2008 M7.9 Wenchuan earthquake in Sichuan, China, some wells in Taiwan, ~2000 km away from the epicenter, documented groundwater oscillations that occurred together not only with Rayleigh waves, but also with S waves and Love waves which are not supposed to have any volumetric strain. These wells (Fig. 5.11), that documented the groundwater-level changes at 1 Hz, are referenced to the same Global Positioning System (GPS) time signal received by the broadband seismometers. Thus the water-level records from these wells may be directly compared with the broadband seismograms from nearby stations, making this dataset excellent and unique for understanding the interaction between seismic waves and water level.

In order to show the association between groundwater oscillations and S and Love waves, we first identify the different seismic phases appearing on the broadband seismograms from a nearby broadband station TPUB (Fig. 5.11), chosen for its average distance to the three wells. Figure 5.12e shows dispersed Love waves appearing on the transverse component; long-period (~50-s) waves, with amplitude ~1 mm/s and a group velocity of 3.9 km/s, first arrived at ~490 s; shorter period

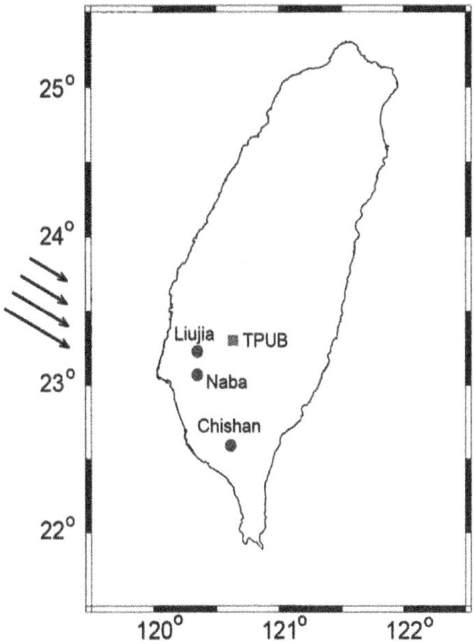

Fig. 5.11 Locations of three monitoring wells (*blue circles*) that documented the water-level changes during the Wenchuan earthquake at a recording rate of 1 Hz and a nearby broadband seismographic station (*red square*). *Arrows* on the *left* show the direction of the seismic waves from the Wenchuan earthquake (From Wang et al., 2009)

(~20-s) waves, with a peak particle velocity of 6 mm/s and a group velocity of 3.2 km/s, arrived at ~600 s after the earthquake. Similarly, Fig. 5.12d,f show dispersed Rayleigh waves appearing on the radial and vertical components; long-period Rayleigh waves first appeared at ~500 s and shorter period waves with larger amplitude arrived at ~650 s. Superimposed on the long-period surface waves are small amplitude (~0.5 mm/s) ~10-s S waves which arrived at ~440 s after the earthquake. The predicted arrival time of S waves at TPUB, according to the IASPEI model, is 442 s.

The water level in the Chishan and the Naba wells (Fig. 5.12a, b) began to oscillate at ~450 s after the earthquake with small amplitudes (~3 mm) and ~10-s period, followed at ~500 s by a rapid decrease of water level of ~1 cm (see inset of Fig. 5.12a), which in turn was followed by oscillations of ~50-s period with greater amplitudes (~1 cm). Much stronger oscillations, with amplitudes up to ~5 cm, started at ~650 s. A sustained decrease of ~3 mm appeared in the Naba well. Comparison between these water-level records with the seismograms shows that the ~50-s oscillations of the water level that started at ~500 s and the ~20-s oscillations that started at ~650 s may be associated with Rayleigh waves, as expected from earlier studies (Cooper et al., 1965; Liu et al., 1989; Brodsky et al., 2003). However, the small water-level oscillations that started at ~450 s coincide in time with the arrival of S-waves, and those that occurred at ~600 s coincide with the arrival of Love waves (Wang et al., 2009). Thus, while the major water-level responses in the Chishan and the Naba wells are consistent with general assumption that water level in the far field are caused by Rayleigh waves (Cooper et al., 1965; Liu et al., 1989), the smaller amplitude oscillations that occurred before Rayleigh waves are not.

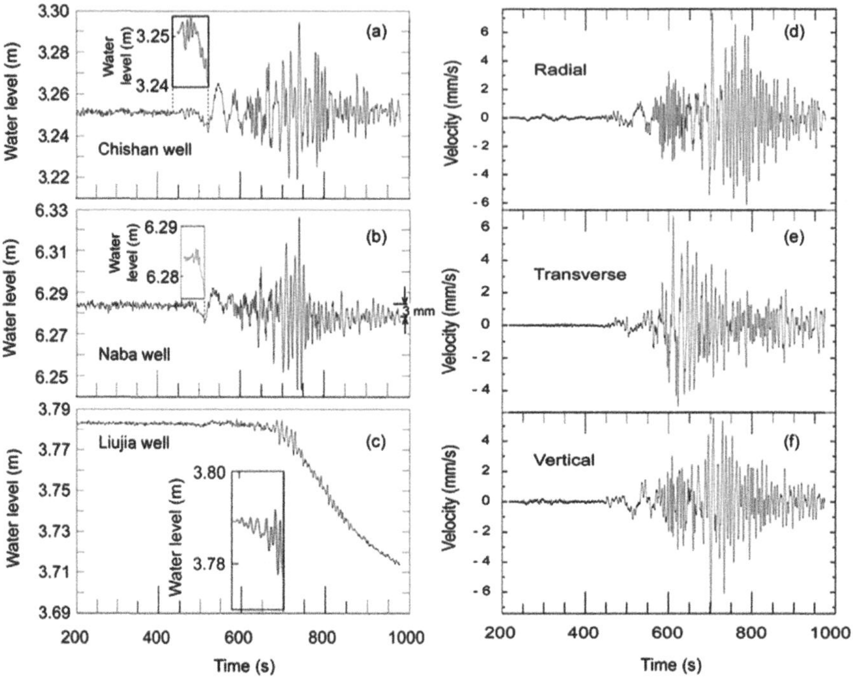

Fig. 5.12 Water level responses to the Wenchuan earthquake in (**a**) the Chishan well, (**b**) the Naba well, and (**c**) the Liujia well. Markings on the vertical axis show elevations above sea-level. Insets show magnified views of the records 50–120 s prior to the occurrence of major responses; a low pass filter of 0.15 Hz was applied to these records to rid off the high-frequency background noise. Broadband seismograms of particle velocity at the TPUB station in (**d**) the radial direction, (**e**) the transverse direction, and (**f**) the vertical direction (courtesy of Data Management Center of the Institute of Earth Sciences, Academia Sinica, Taiwan). Time zero refers to the time of the Wenchuan earthquake (From Wang et al., 2009)

The association of water-level changes with Love waves and S waves could be due to an anisotropic poroelastic effect (Wang, 2000; Brodsky et al., 2003); but a demonstration of this effect, which requires a knowledge of the three dimensional strain tensors at the wells, is lacking.

In the Liujia well, the ~50-s period oscillation is barely discernable (Fig. 5.12c). At ~600 s when the large amplitude ~20-s Love waves arrived, a train of ~20-s oscillations occurred with increasing amplitude (up to ~1 cm) and the water level began to decline (see inset of Fig. 5.12c). This is followed at ~650 s by a pronounced decrease of water level that lasted long after the oscillations had stopped, with a total decrease of ~ 9 cm. The sequence of events suggests that the sustained decline of water level may have been initiated by the train of water-level oscillations.

The small water-level oscillations associated with S and Love waves must be caused by groundwater flow into and out of the aquifer. Wang et al. (2009) showed that the groundwater flow has velocity large enough to break up colloidal aggregates in the pore water thus to enhance aquifer permeability, suggesting that the small water-level oscillations may have facilitated the occurrence of the sustained decline of water level in the Liujia well.

The strikingly different responses of the water level between the Liujia well and the other two wells, all at about the same distance from the Wenchuan earthquake, shows that the boundary is not sharp between the sustained groundwater-level response, supposedly occurring in the intermediate field, and the oscillation without sustained changes, supposedly occurring in the far field, and that there may be a broad transition in which both types of response may occur together. One reason such different responses may occur together is the different transmissivity of the aquifers intersected by these wells. While the screened aquifer at the Liujia well consists of layered fine sands, silt, mud and clay, those at the other two wells consist of either gravels or uniform sands (Wang et al., 2009). In the absence of accurate well tests, we may infer from the well logs that the transmissivity (the product of the hydraulic conductivity of the aquifer and aquifer thickness) is lowest at the Liujia and highest at the Chishan well, and intermediate at the Naba well. The different transmissivities may account for the different oscillation characteristics of the wells, i.e., vigorous coseismic oscillation in the Chishan and Naba wells but barely discernable in the Liujia well until the arrival of the strong 20-s Love waves. The different transmissivity may even account for the absence of a sustained change of groundwater level in the Chishan well, but a pronounced change in the Liujia well. High transmissivity promotes uniform pore pressure; thus the probability for an enhanced permeability channel to intersect a reservoir of different pressure is low. On the other hand, poor transmissivity can support heterogeneous pore pressures in close proximity, and thus the probability for an enhanced permeability channel to intersect a reservoir of different pressure is higher.

5.6 Pore-Pressure Changes on the Sea Floor

So far, our discussion has been focused on groundwater-level changes on land. But pore-pressure changes during earthquakes have also been documented beneath the sea floor. Due to the sparse instrumentation on the sea floor, however, the available data are far fewer in number than the land-based data and are further localized in specific tectonic areas.

An earthquake swarm on the Juan de Fuca Ridge, that occurred on June 8, 1999, and lasted over a span of more than two months (Fig. 5.13c), caused pore-pressure transients (Fig. 5.13a, b) that were recorded in Ocean Drilling Program (ODP) boreholes on the eastern flank of the ridge, 25–100 km from the epicenter. These transients were characterized by a rapid coseismic rise in pressure, followed by a continuing slower rise to a peak, and then a much slower decay (Fig. 5.13b). Thus the time-history of these pore-pressure transients is similar to the sustained groundwater level changes described in Sect. 5.3. As noted by Davis et al. (2001), the pore-pressure transients occurred only during the first earthquake, but not during

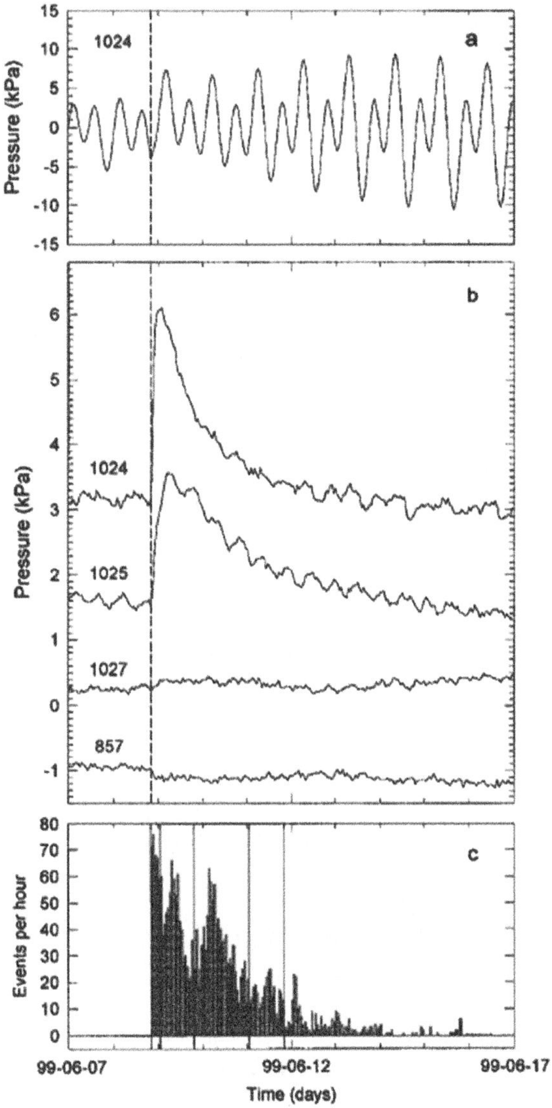

Fig. 5.13 (**a**) Raw formation pressure record from ODP Site 1024C at the time of the June 1999 earthquake swarm along the Endeavour ridge segment. (**b**) Pore-pressure records from this site and several other ODP Sites after the responses to tidal, barometric, and oceanographic loading were removed. (**c**) Earthquakes recorded at onshore seismic stations (*vertical lines*) and histogram of the number of events detected for the same time (From Davis et al., 2001)

the latter earthquakes, even though several of the latter shocks were of greater magnitude than the first one. Davis et al. (2001) explained the absence of pore-pressure transients during the latter earthquakes by suggesting that the pore-pressure change was caused by a spreading event *at the time* of the first earthquake, but most deformation during the latter events occurred aseismically; thus these earthquakes were merely the seismic expression of a much bigger tectonic event, not the cause of the pore-pressure transients.

Here we offer an alternative interpretation of the non-responsiveness of pore pressure to the latter earthquakes. According to the enhanced permeability model, the first earthquake opened some permeable passageways between the ODP site and local high-pressure sources, which caused the observed increase in pore pressure at the borehole. In order to cause a second increase in pore pressure, sufficient time must pass to allow the fluid passageways to seal and the high-pressure sources to re-pressurize. Based on data from Iceland, Claesson et al. (2007) suggested that a postseismic recovery of permeability took about 2 years. Similar repose times were documented for earthquake triggering of mud volcanoes (Mellors et al., 2007; Manga et al., 2009). Thus, for the case of the earthquake swarm on the Juan de Fuca Ridge, there may simply be insufficient time between the first and the subsequent earthquakes to allow the permeable channels to re-seal and the local sources to re-pressurize. Hence, after the pore pressure transient induced by the first earthquake of the swarm, no more pore-pressure transients were possible during the remaining two-month span of the swarm.

5.7 Postseismic Groundwater Recession

Following the step-like changes in the groundwater level in the near field, postseismic flow in the aquifer leads to a new equilibrium state. Here we term this time-dependent recovery of the groundwater level the 'postseismic groundwater recession'. We show in the following that an analysis of the postseismic groundwater recession may provide new information on the characteristics of the aquifers at a spatial scale which is otherwise unattainable.

5.7.1 Recession Analysis

For the postseismic recession analysis, we select the records from wells that are free from significant anthropogenic disturbances. We then determine, for each groundwater level record, a postseismic equilibrium level from the record. This is subtracted from the raw record; the difference gives the time-history of the postseismic groundwater level.

By plotting the logarithm of the postseismic residuals of the groundwater level, denoted as h, against time we found that the relations between $log\, h$ and t are linear after a sufficient lapse of time (Fig. 5.14); i.e.,

$$log\, h = a - bt \tag{5.3}$$

similar to the relation in the recession of stream flow discussed in Chap. 4. Here a and b are empirical constants to be evaluated from least-square fit to well data; a minus sign is placed in front of the empirical constant b so that b itself is positive. The constant b, to be distinguished from the empirical constant c estimated from the recession of stream discharge discussed in Chap. 4, may be estimated from the

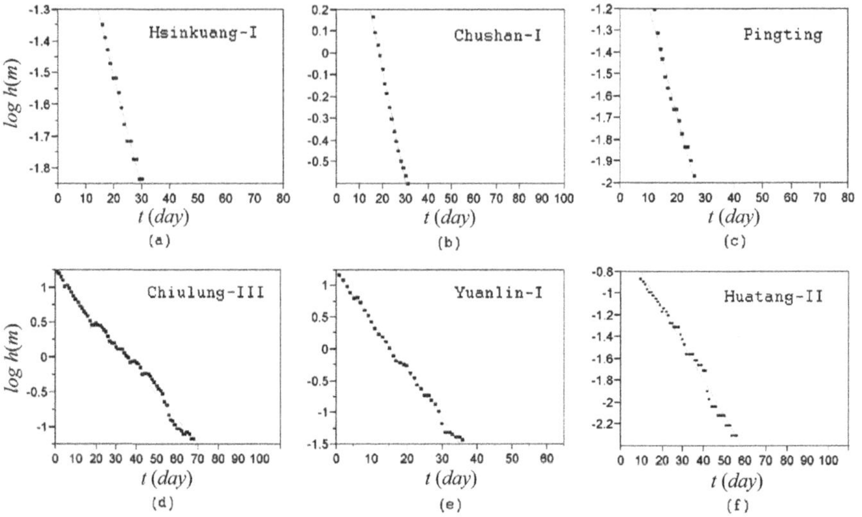

Fig. 5.14 Diagrams showing logarithmic of postseismic residual hydraulic head plotted against time for six wells. *Dots* are data; *straight lines* are least-square linear fits to data. R-squares of linear fits are listed in Table 5.2 (From Wang et al., 2004)

recession of the groundwater level. As shown by the next section, *b* and *c* are physically equivalent, even though they are evaluated from different hydrologic data. Thus the study of earthquake hydrology provides another means to estimate the hydrogeologic parameters of aquifers.

Table 5.2 lists the values for *b* and the square of the correlation coefficient, R^2, determined from the least square fit of the data for the most common type of postseismic changes in the groundwater level (Fig. 5.1b). For other types of postseismic changes, the readers are referred to Wang et al. (2004). We again define a characteristic time as

Table 5.2 List of the values for b, τ for aquifers, and R^2, for the least-square fit with Eq. (5.5) to data for the postseismic change of groundwater level at several stations

		Aquifer I	Aquifer II	Aquifer III	R^2
Chiulung	b			$4.0 \times 10^{-7}\ s^{-1}$	0.99
	τ			2.5×10^{6} s	
Hsihu	b			$3.6 \times 10^{-7}\ s^{-1}$	0.99
	τ			2.8×10^{6} s	
Huatang	b		$3.8 \times 10^{-7}\ s^{-1}$		0.99
	τ		2.6×10^{6} s		
Kuoshen	b		$9.7 \times 10^{-7}\ s^{-1}$		0.98
	τ		1.0×10^{6} s		
Yuanlin	b	$8.8 \times 10^{-7}\ s^{-1}$	$9.3 \times 10^{-7}\ s^{-1}$		1.0*, 0.95^
	τ	1.1×10^{6} s	1.1×10^{6} s		

*Aquifer I; ^Aquifer II.

$$\tau \equiv 1/b. \tag{5.4}$$

As shown in Table 5.2, the characteristic times range from 1 to 3×10^6 s. It is interesting to note that the result from this analysis is similar to that obtained from the analysis of the postseismic baseflow recession in the last chapter (Sect. 4.3.1, Table 4.1).

5.7.2 Interpretation of the Postseismic Recession

As in the streamflow analysis, we use a simple model to simulate the postseismic groundwater flow. Again we approximate the aquifer by a one-dimensional layer (Fig. 4.5a), extending from a local groundwater divide ($x = 0$) to a local discharge or recharge area ($x = L$). We assume that the entire aquifer consolidates at the time of the earthquake, $t = 0$, and releases an amount of water A_0 per unit volume. With the boundary conditions of no-flow at $x = 0$ and $h = 0$ at $x = L$, the solution of the flow equation (4.2) is (Appendix B.10):

$$h(x,t) = \frac{4A_o}{\pi S_s} \sum_{r=1}^{\infty} \frac{(-1)^{r+1}}{(2r-1)} \cos \frac{(2r-1)\pi x}{2L} \exp\left[-(2r-1)^2 \frac{t}{\tau}\right]. \tag{5.5}$$

Given the characteristic time τ determined from the recession analysis (Table 5.2) and the specific storage S_s of 10^{-4} m^{-1} from well tests (Tyan et al., 1996), we may compare the model prediction with the postseismic time history of the groundwater level change. As an example, we show in Fig. 5.15 the predictions from the model for different values of x/L against the postseismic groundwater level data at Yuanlin I well, where x is the position of the well, which is not known *a priori* and must be determined from model fitting. The curve for $x/L = 0.5$ shows an excellent fit to the field data (Wang et al., 2004), suggesting that the Yuanlin station may be situated near the mid-point of an aquifer between the local groundwater divide and the local discharge location (Fig. 5.15).

From Eq. (5.3) and the above result, we have, after a sufficient lapse of time

$$b = -\frac{\partial \log h}{\partial t} \approx \frac{\pi^2 D}{4L^2}. \tag{5.6}$$

where $D = K/S_s$ (Eq. B.21). From the values of b in Table 5.2 and D from well tests, we may estimate the characteristic length L of the aquifer from (5.6). Using an average value of $D \sim 10$ m^2/s from well tests for the confined aquifers in the Choshui River fans (Lee and Wu, 1996; Kester and Ouyang, 1996; Tyan et al., 1996) and $b \sim 10^{-6}$ s^{-1} (Table 5.2), we obtain $L \sim 5000$ m for the confined aquifers. Thus the step-like rise of groundwater level, as documented by wells installed in the Choshui River fan, was dissipated by groundwater flow through subhorizontal aquifers with an average characteristic length of 5 km. This characteristic length is considerably smaller than the length of confined aquifers indicated on a geologic cross-section

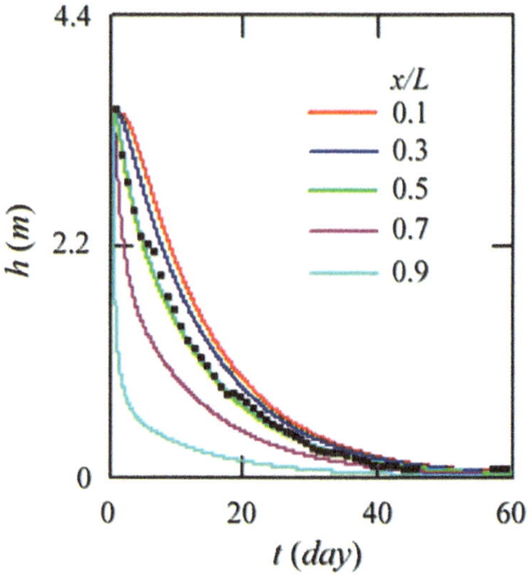

Fig. 5.15 Diagram showing documented time-history of postseismic groundwater level change at Yuanlin I well. *Black dots* show data points. Model prediction of postseismic changes of h are given by *colored curves* for several values of x/L. Excellent fit between data and curve occurs for $x/L = 0.5$ (From Wang et al., 2004)

(e.g., Fig. 5.3). This difference suggests that the actual geologic structure of the aquifers in the alluvial fan may be much more heterogeneous and complex than that implied by the simplified geologic cross-section (Fig. 5.3) reconstructed from limited number of well logs.

Finally, the postseismic recession of groundwater level (i.e., a corresponding decrease in pore pressure) implies a postseismic consolidation and thus a reduction of porosity for most aquifers on the Choshui River fan. Wang et al. (2001) estimated a porosity reduction of 10^{-3} to 10^{-4}. Although this is too small to be measured with the current techniques, the accumulated porosity reduction by this process may be significant over geologic time. Assuming a recurrence time of 10^2 yr for large earthquakes and an average porosity reduction of 10^{-4} m^3/m^3 per large earthquake, the accumulated porosity reduction would be $\sim$10% over 10^5 yr – a brief moment in geological time. Thus earthquake-induced consolidation may play an important role on sediment compaction and in driving fluid flow in sedimentary basins.

5.8 Concluding Remarks

While the variety in the water-level response to earthquakes may at first glance appear bewildering, there is a pattern and trend scaled by a decrease of seismic energy density with increasing distance from the ruptured fault. Very close to the ruptured fault, strong shaking may cause pre-existing fractures to open and rock/sediments to dilate, leading to a step-like decrease in water level. In the near field, but outside the immediate neighborhood of the ruptured fault, consolidation of loose sediments may lead to a step-like increase in water level. In fractured crystalline rocks, however, the pattern of water level changes may mimic the co-seismic volumetric strain. In the intermediate field, permeability changes can result in either positive or negative sustained changes in water level. In the far field, sustained changes and oscillations of groundwater level can occur in response to Rayleigh waves, as well as to S- and Love waves.

References

Akita, F., and N. Matsumoto, 2004, Hydrological responses induced by the Tokachi-oki earthquake in 2003 at hot spring well in Hokkaido, *Japan. Geophys. Res. Lett., 31*, doi:10.1029/2004GL020433.

Blanchard, F.G., and P. Byerly, 1935, A study of a well gage as a seismograph, *Bull. Seismo. Soc. Am., 25,* 313–321.

Brodsky, E.E., E. Roeloffs, D. Woodcock, I. Gall, and M. Manga, 2003, A mechanism for sustained groundwater pressure changes induced by distant earthquakes, *J. Geophys. Res., 108*, 2390, doi:10.1029/2002JB002321.

Chia, Y.P., Y.S. Wang, H.P. Wu, J.J. Chiu, and C.W. Liu, 2001, Changes of groundwater level due to the 1999 Chi-Chi earthquake in the Choshui *River* fan in Taiwan, *Bull. Seism. Soc. Am., 91*, 1062–1068.

Chia, Y., J.J. Chiu, Y.H. Jiang, T.P. Lee, Y.M. Wu, and M.J. Horng, 2008, Implications of coseismic groundwater level changes observed at multiple-well monitoring stations, *Geophys. J. Int., 172,* 293–301.

Claesson, L., A. Skelton, C. Graham, and C.-M. Mörth, 2007, The timescale and mechanisms of fault sealing and water-rock interaction after an earthquake, *Geofluids, 7*, 427–440.

Cooper, H.H., J.D. Bredhoeft, I.S. Papdopulos, and R.R. Bennnett, 1965, The response of aquifer-well systems to seismic waves, *J. Geophys. Res., 70*, 3915–3926.

Davis, E.E., K. Wang, R.E. Thomson, K. Becker, and J.F. Cassidy, 2001, An episode of seafloor spreading and associated plate deformation inferred from crustal fluid pressure transients, *J. Geophys. Res., 106*, 21953–21963.

Dobry, R., R.S. Ladd, F.Y. Yokel, R.M. Chung, and D. Powell, 1982, Prediction of pore water pressure buildup and liquefaction of sands during earthquakes by the cyclic strain method, *National Bureau of Standards Building Science Series, 138*, National Bureau of Standards and Technology, Gaithersburg, Md., pp. 150.

Elkhoury, J.E., E.E. Brodsky, and D.C. Agnew, 2006, Seismic waves increase permeability, *Nature, 411*, 1135–1138.

Ge, S., and C. Stover, 2000, Hydrodynamic response to strike- and dip-slip faulting in a half space, *J. Geophys. Res., 105*, 25513–25524.

Hazirbaba, K., and E.M. Rathje, 2004, A comparison between in situ and laboratory measurements of pore water pressure generation, in *13th World Conference on Earthquake Engineering, paper no. 1220*, Vancouver.

Holzer, T.L., and T.L. Youd, 2007, Liquefaction, ground oscillation, and soil deformation at the Wildlife Array, California, *Bull. Seis. Soc. Am., 97*, 961–976.

Hsieh, P., J. Bredehoeft, and J. Farr, 1987, Determination of aquifer permeability from earthtide analysis, *Water Resour. Res, 23*, 1824–1832.

Hsu, C.C., and M. Vucetic, 2004, Volumetric threshold shear strain for cyclic settlement, *J. Geotech. Geoenviron. Eng., 130*, 58–70.

Igarashi, G., and H. Wakita, 1991, Tidal responses and earthquake-related changes in the water level of deep wells, *J. Geophys. Res., 96*, 4269–4278.

Igarashi, G., and H. Wakita, 1995 Geochemical and hydrological observations for earthquake prediction in Japan, *J. Phys. Earth, 43*, 585–598.

Institute of Geophysics–CAS (China Earthquake Administration), 1976, China earthquake catalog, 500 p, Washington, DC: Center for Chinese Research Materials (in Chinese).

Ishihara, K., 1996, *Soil Behavior in Earthquake Geotechnics*, 350 pp, Oxford: Clarendon Press.

Itaba, S., and N. Koizumi, 2007, Earthquake-related changes in groundwater levels at the Dogo hot spring, Japan, *Pure Appl. Geophys., 164*, 2397–2410.

Jonsson, S., P. Segall, R. Pedersen, and G. Bjornsson, 2003, Postearthquake ground movements correlated to pore-pressure transients, *Nature, 424*, 179–183.

Kester, L.C., and S. Ouyang, 1996, A first investigation of the groundwater and its equilibrium in the Yunlin district. In: *Conference on Groundwater and Hydrology of the* Choshui alluvial fan, *Taiwan*, Water Resources Bureau, 181–206 (in Chinese).

King, C.-Y., S. Azuma, G. Igarashi, M. Ohno, H. Saito, and H. Wakita, 1999, Earthquake-related water-level changes at 16 closely clustered wells in Tono, central Japan, *J. Geophys. Res.,104*, 13073–13082.

Kitagawa, Y., N. Koizumi, M. Takahashi, N. Matsumoyo, and T. Sato, 2006, Changes in water levels or pressures associated with the 2004 earthquake off the west coast of northern Sumatra (M9.0), *Earth Planets Space, 58*, 173–179.

Koizumi, N., W.-C. Lai, Y. Kitagawa, and Y. Matsumoto, 2004, Comment on "Coseismic hydrological changes associated with dislocation of the September 21, 1999 Chichi earthquake, Taiwan" by Min Lee et al., *Geophys. Res. Lett., 31*, L13603, doi:10.1029/2004GL019897

Kono, Y., and T. Yanagidani, 2006, Broadband hydroseismograms observed by closed borehole wells in the Makioka mine, central Japan: Response of pore pressure to seismic waves from 0.05 to 2 Hz, *J. Geophys. Res., 111*, B03410 doi:10.1029/2005JB003656.

Lee, C.S., and C.L. Wu, 1996, Pumping tests of the Choshuichi alluvial fan, in *Conference on Groundwater and Hydrology of the* Choshui *River* Alluvial Fan, *Taiwan*, Water Resources Bureau, 165–179 (in Chinese).

Liu, L.B., E. Roeloffs, and X.Y. Zheng, 1989, Seismically induced water level oscillations in the Wali well, Beijing, China, *J. Geophys. Res., 94*, 9453–9462.

Luong, M.P., 1980, Stress-strain aspects of cohesionless soils under cyclic and transient loading. In: G.N. Pande, and O.C. Zienkiewicz (eds.), *Proc. Intern. Symp. Soils under Cyclic and Transient Loading*, p. 315–324, Rotterdam, Netherlands: A.A. Balkema.

Manga, M., and C.-Y. Wang, 2007, Earthquake hydrology, In: H. Kanamori (ed.), *Treatise on Geophysics, 4*, Elsevier, Ch. In: *Treatise on Geophysics*, G. Schubert editor, Vol. 4, 293–320.

Manga, M., M. Brumm, and M.L. Rudolph, 2009, Earthquake triggering of mud volcanoes. *Mar. Pet. Geol., 26,* 1785-1798.

Matsumoto, N., G. Kitagawa, and E.A. Roeloffs, 2003, Hydrological response to earthquakes in the Haibara well, central Japan – I. Water level changes revealed using state space decomposition of atmospheric pressure, rainfall and tidal responses, *Geophys. J. Int., 155*, 885–898.

Mellors, R., D. Kilb, A. Aliyev, A. Gasanov, and G.Yetirmishli, 2007, Correlations between earthquakes and large mud volcano eruptions, *J. Geophys. Res., 112*, B04304.

Montgomery, D.R., and M. Manga, 2003, Streamflow and water well responses to earthquakes, *Science, 300*, 2047–2049.

National Research Council, 1985, *Liquefaction of Soils during Earthquakes*, pp. 240, Washington, DC: National Academy Press.

Quilty, E., and E. Roeloffs, 1997, Water level changes in response to the December 20, 1994, M4.7 earthquake near Parkfield, California, *Bull. Seismol. Soc. Am., 87*, 310–317.

Roeloffs, E.A., 1996, Poroelastic methods in the study of earthquake-related hydrologic phenomena. In: R. Dmowska (ed.), *Advances in Geophysics*, San Diego: Academic Press.

Roeloffs, E.A., 1998, Persistent water level changes in a well near Parkfield, California, due to local and distant earthquakes, *J. Geophys. Res., 103*, 869–889.

Roeloffs, E.A., M. Sneed, D.L. Galloway, M.L. Sorey, C.D. Farrar, J.F. Howle, and J. Hughes, 2003, Water-level changes induced by local and distant earthquakes at Long Valley caldera, California, *J. Volcan. Geotherm. Res., 127*, 269–303.

Rojstaczer, S., S. Wolf, and R. Michel, 1995, Permeability enhancement in the shallow crust as a cause of earthquake-induced hydrological changes, *Nature, 373*, 237–239.

Sato, T., N. Matsumoto, Y. Kitagawa, N. Koizumi, M. Takahashi, Y. Kuwahara, H. Ito, A. Cho, T. Satoh, K. Ozawa, and S. Tasaka, 2004, Changes in water level associated with the 2003 Tokachi-oki earthquake, *Earth Planets Space, 56*, 395–400.

Seed, H. B., and K. L. Lee, 1966, Liquefaction of saturated sands during cyclic loading, *J. Soil Mech. Found. Div., 92*, 105–134.

Sil, S., and J.T. Freymueller, 2006, Well water level changes in Fairbanks, Alaska, due to the great Sumatra-Andaman earthquake, *Earth Planets Space, 58*, 181–184.

Tyan, C.L., Y.M. Chang, W.K. Lin, and M.K. Tsai, 1996, The brief introduction to the groundwater hydrology of Choshui River Alluvial fan, in *Conference on Groundwater and Hydrology of the* Choshui *River* Alluvial Fan, *Taiwan*, Water Resources Bureau, 207–221 (in Chinese).

Vucetic, M., 1994, Cyclic threshold of shear strains in soils, *J. Geotech. Eng., 120*, 2208–2228.

Wakita, H., 1975, Water wells as possible indicators of tectonic strain, *Science, 189*, 553–555.

Wang, C.-H., C.-Y. Wang, C.-H. Kuo, and W.-F. Chen, 2005, Some isotopic and hydrological changes associated with the 1999 Chi-Chi earthquake, Taiwan, *The Island Arc, 14*, 37–54.

Wang, C.-Y., 2007, Liquefaction beyond the near field, *Seism. Res. Lett., 78*, 512–517.

Wang, C.-Y., L.H. Cheng, C.V. Chin, and S.B. Yu, 2001, Coseismic hydrologic response of an alluvial fan to the 1999 Chi-Chi earthquake, Taiwan, *Geology, 29*, 831–834.

Wang, C.-Y., D.S. Dreger, C.-H. Wang, D. Mayeri, and J.G. Berryman, 2003, Field relations among coseismic ground motion, water level change and liquefaction for the 1999 Chi-Chi (M_W= 7.5) earthquake, Taiwan, *Geophys. Res. Lett., 30*, doi:10.1029/2003GL017601.

Wang, C.-Y., C.-H. Wang, and C.-H. Kuo, 2004, Temporal change in groundwater level following the 1999 (M_W = 7.5) Chi-Chi earthquake (1999), Taiwan, *Geofluids, 4*, 210–220.

Wang, C.-Y., and Y. Chia, 2008, Mechanism of water level changes during earthquakes: Near field versus intermediate field, *Geophys. Res. Lett., 35*, L12402, doi:10.1029/2008GL034227.

Wang, C.-Y., Y. Chia, P.-L. Wang, and D. Dreger, 2009, Role of S waves and Love waves in coseismic permeability enhancement, *Geophys. Res. Lett.*, 36, L09404, doi:10.1029/2009GL037330.

Wang, H., 2000, *Theory of Linear Poroelasticity*, Princeton, NJ: Princeton University.

Wells, D.L., and K.J. Coppersmith, 1994, New empirical relationships among magnitude, rupture length, rupture width, rupture area, and surface displacement, *Bull. Seis. Soc. Am., 84*, 974–1002.

Yoshimi, Y., and H. Oh-Oka, 1975, Influence of degree of shear stress reversal on the liquefaction potential of saturated sand, *Soils and Foundations* (Japan), *15*, 27–40.

Chapter 6
Temperature and Composition Changes

Contents

6.1 Introduction

Changes in the temperature, odor, and taste of groundwater are probably among the earliest reported changes following earthquakes. These changes may be expected, not only because earthquake-induced groundwater flow is effective in transporting heat and solutes, but because significant amounts of frictional heat could be generated along the displaced fault, which may raise groundwater temperature. Progress in our understanding of these processes, however, has been slow, largely because relevant quantitative data are scarce. Systematic measurements of earthquake-induced changes in temperature and composition started only in the late twentieth century. Continuous recording of temperature has become available in a limited number of wells and springs. Composition records are even fewer because most measurements require discrete sampling of water and expensive and time-consuming laboratory analysis. Another complication for the temperature and composition data is that the measured values may depend strongly on the proximity of the point of measurement from hydraulically conductive fractures (e.g., Barton et al., 1995). Because temperature measurements are normally made with one or two probes at fixed depths, often

C.-Y. Wang, M. Manga, *Earthquakes and Water*, Lecture Notes in Earth Sciences 114, DOI 10.1007/978-3-642-00810-8_6, © Springer-Verlag Berlin Heidelberg 2010

without clear reference to conductive fractures, and composition measurements are normally made on sampled well waters from depth, again without clear reference to conductive fractures, they may be recording different features of the postseismic changes, and it is not surprising that the data are sometimes difficult to interpret. In this chapter we first summarize some significant observations on earthquake-induced temperature changes in hot springs, wells, and marine geothermal systems, we then summarize those on earthquake-induced compositional changes. After the summary of the observations, we discuss the mechanisms for the changes, both those proposed by previous authors and those proposed here for the first time.

6.2 Earthquake-Induced Change in Groundwater Temperature

6.2.1 Hot Springs

Mogi et al. (1989) described the temperature change in a hot spring in the north-eastern part of the Izu Peninsula, Japan. Four temperature probes were installed in the well, but only the topmost probe's data were reported. Figure 6.1 shows that, when there are no earthquakes, the temperature of the well water falls gradually and linearly with time; at the time of earthquakes, on the other hand, temperature rises in a stepwise pattern. Mogi et al. (1989) interpreted the gradual decline of temperature during normal times to indicate a decrease in the amount of geothermal water in the hot spring as a result of ongoing precipitation of obstacles in underground passageways, slowly blocking the flow of the geothermal water. When a fairly strong earthquake occurs, the seismic waves dislodge the obstacles, and the flow of the geothermal water suddenly increases and temperature suddenly rises.

Mogi et al. (1989) further plotted the epicentral distance of earthquake versus magnitude, for earthquakes of magnitude between 2 and 5 that occurred within a radius of 600 km of this artesian spring (Fig. 6.2). Earthquakes in which coseismic changes occurred are shown by solid circles and ones in which no such changes occurred are shown by open circles. For earthquakes whose epicentral distance was over 20 km, Mogi et al. (1989) found that the solid and open circles can be separated by a magnitude-versus-logarithm of distance threshold indicated by the solid line in Fig. 6.2. An interpretation of this line is given later in the discussion of mechanism (Sect. 6.2.4). It is interesting to note that this line is similar to the thresholds that delimit the occurrences of earthquake-induced liquefaction (Chap. 2; Fig. 6.6), streamflow increase (Chap. 4; Fig. 4.9) and sustained groundwater level change (Chap. 5; Fig. 5.6). An integrated interpretation of the similarities and differences among these threshold relations is given in Chap. 10.

At distances $\leq$ 20 km and $M \leq 4$, the open and solid circles are mixed. Mogi et al. (1989) interpreted these to represent a mix of foreshocks and earthquake swarms. The earthquakes for which changes were regarded as precursory are marked in Fig. 6.2 by ringed circles. This part of the diagram will be discussed in Chap. 9 on earthquake precursors.

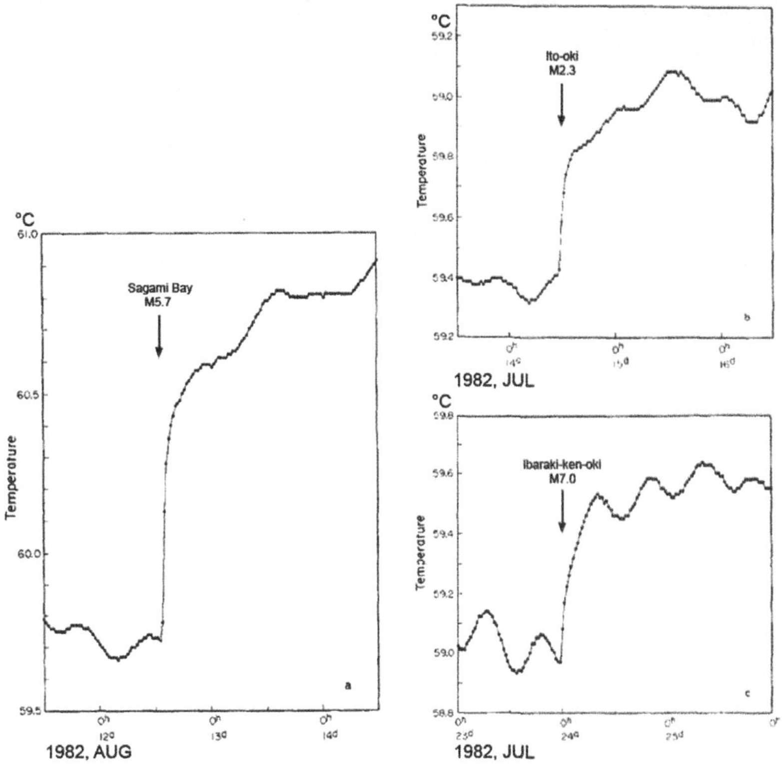

Fig. 6.1 Examples of coseismic changes in the temperature at Usami No. 24 hot spring: (**a**) the Sagami Bay earthquake (August 1982) of $M = 5.7$; (**b**) the Ito-oki earthquake (July 1982) of $M = 2.3$; (**c**) the Ibaraki-ken-oki earthquake (July 1982) of $M = 7.0$ (From Mogi et al., 1989)

6.2.2 Wells

Changes of temperature in response to earthquakes have been documented in many wells since historical time (e.g., Ma et al., 1990). As an example, we discuss the earthquake-induced temperature changes documented in a well in the city of Tangshan, China. Simultaneous and continuous measurements of temperature and groundwater level have been carried out in this well since 2001, and have documented coseismic temperature changes and well water oscillations in response to many distant earthquakes (Shi et al., 2007). Temperature is measured by a high-resolution temperature probe (10^{-4} °C precision is claimed) installed 125 m beneath the well head. Figure 6.3 shows the water-level oscillations and temperature changes recorded during 12 earthquakes (note that some traces contain two earthquakes). As soon as the surface waves arrive, water level in the well oscillates and temperature

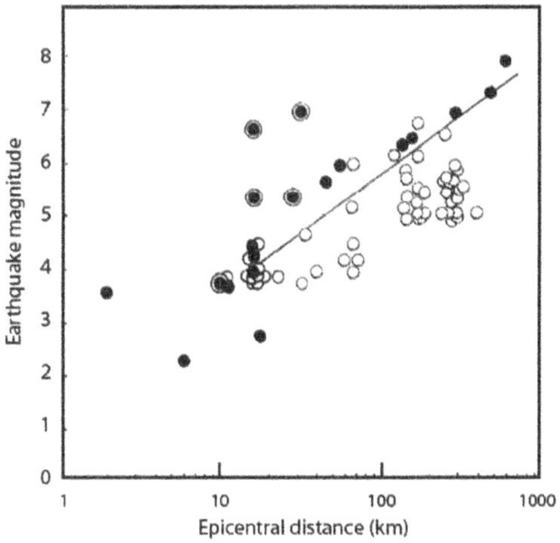

Fig. 6.2 *Solid circle* – coseismic change; *ringed solid circle* – precursory change; *open circle* – no change (From Mogi et al., 1989)

drops until the oscillations stop. Temperature always drops independent of the orientation of the causal fault, the distance from the hypocenter, and the magnitude of the seismic events. The rate of temperature drop is generally rapid; it begins when the seismic waves first arrive, reaches a maximum amplitude from 0.001 to

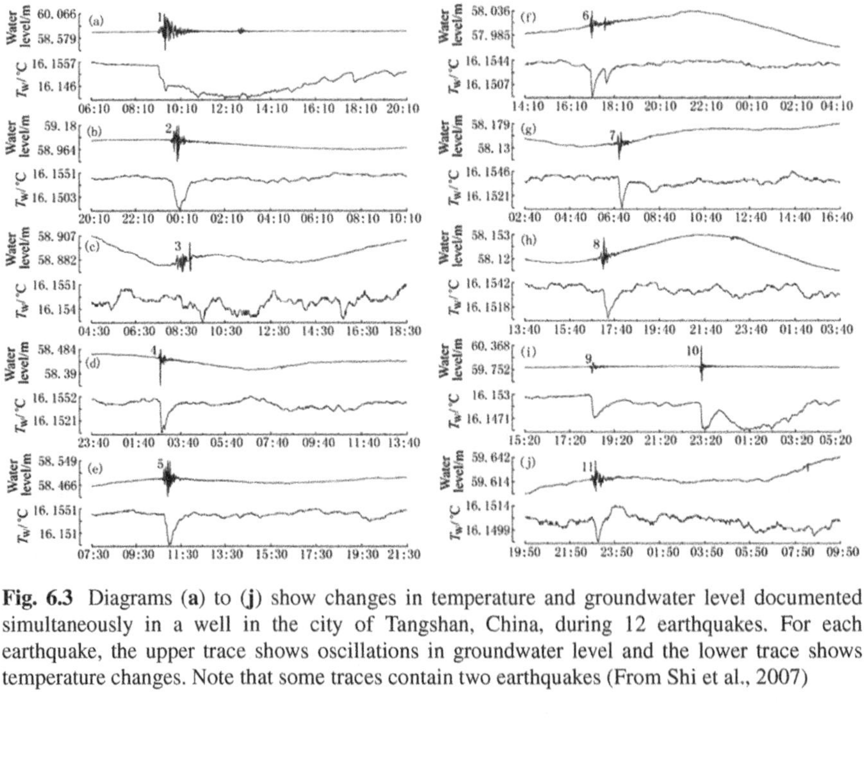

Fig. 6.3 Diagrams (**a**) to (**j**) show changes in temperature and groundwater level documented simultaneously in a well in the city of Tangshan, China, during 12 earthquakes. For each earthquake, the upper trace shows oscillations in groundwater level and the lower trace shows temperature changes. Note that some traces contain two earthquakes (From Shi et al., 2007)

0.01°C when the well-water oscillation reaches peak amplitude from several centimeters to about one meter, and then returns slowly to the original temperature. The temperature decrease generally takes 10~20 min., and the restoring process takes from 0.5 to several hours. During the great Sumatra earthquake of 2004, many stations in China also observed coseismic temperature changes and most of these show temperature drops (Chap. 8 in Department of Monitoring and Prediction of China Earthquake Administration, 2005).

6.2.3 Marine Hydrothermal Systems

Advances in the past three decades in monitoring geologic systems along and on the flanks of mid-oceanic ridges have revealed abundant evidence of temperature changes during and/or after earthquakes. Some of these were responses to earthquake swarms located directly below the vents (Sohn et al., 1998; Baker et al., 1999), some were responses to earthquakes along adjacent spreading centers (Dziak et al., 2003), and some others were responses to ridge-flank earthquake swarms with epicenter distances up to 50 km away (Johnson et al., 2000, 2001). As shown in the next section, several types of temperature changes have been documented.

6.2.3.1 Temperature Change in Hydrothermal Vents on Mid-Oceanic Ridges

Temperature of hydrothermal vents along mid-oceanic ridges has been found to change in response to local and distant earthquakes (Sohn et al. 1998, 1999; Johnson et al. 2000, 2001, 2006; Dziak et al., 2003). As an example, we discuss the results of some recent studies of temperature changes measured in the vents along the Juan de Fuca Ridge in response to local and distant earthquake swarms.

On June 8, 1999, an earthquake swarm occurred beneath a segment of the ridge and lasted about a week (Johnson et al., 2001). A thermistor array, deployed before the earthquake swarm within a low temperature vent system on the Juan de Fuca Ridge, 7.5 km away from the earthquake swarm, recorded widespread increases of temperature. In Fig. 6.4 (from Johnson et al., 2000), the gray band shows the occurrence of the earthquake swarm, the upper curve shows the temperature from a thermistor directly in a vent, and the lower curve shows the temperature of the axial valley bottom water. The earthquake swarm produced a slow increase in vent temperature 8 days after the initiation of the swarm. All monitored vents within the axial valley responded similarly, with delayed responses varying from a few days to a month and the net heat flux increased by a factor of ten (Johnson et al., 2001).

During June 1–7, 2000, another earthquake swarm, with 170 earthquakes and a mainshock of M_W6.2, occurred on the western Blanco Transform Fault. Two temperature probes, located in hydrothermal vents in an adjacent spreading center on the Juan de Fuca Ridge, ~39 km away from the earthquake swarm, both registered

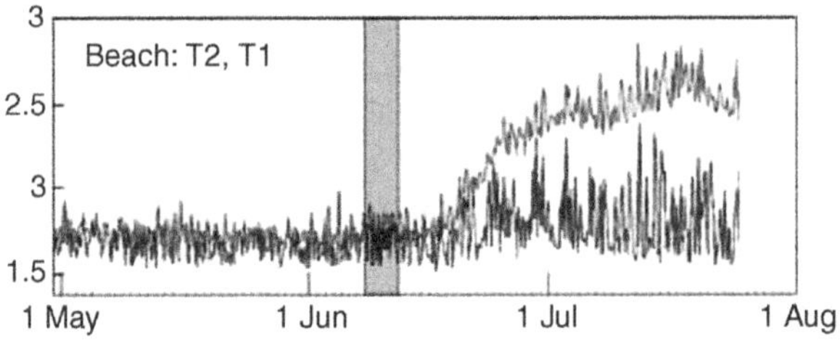

Fig. 6.4 Temperature records from a site on the Endeavour axial valley of the Juan de Fuca Ridge. The June 8–15, 1999, earthquake swarm is marked by the *vertical shaded bar*. *Upper* trace indicate data from thermistors located within the vents; *blue* trace indicate data from thermistors deployed in the adjacent (non-vent) *bottom* water. Vertical axis is temperature in degrees C (From Johnson et al., 2000)

temperature decreases, one occurring over days to weeks while the other changes were coseismic, as shown in the Fig. 6.5. The onset of the temperature decreases was gradual, but accelerated after the occurrence of the earthquake swarm, with a total decrease of more than 20°C, as shown in Fig. 6.5 (Dziak et al. 2003).

The fact that earthquakes can influence sub-surface hydrothermal fluids on the sea floor over significant distances from the epicenters, by either increasing or decreasing temperature and flow rates, implies that fluids within subsurface aquifers beneath the sea floor are frequently 'stirred' tectonically (Dziak et al., 2003).

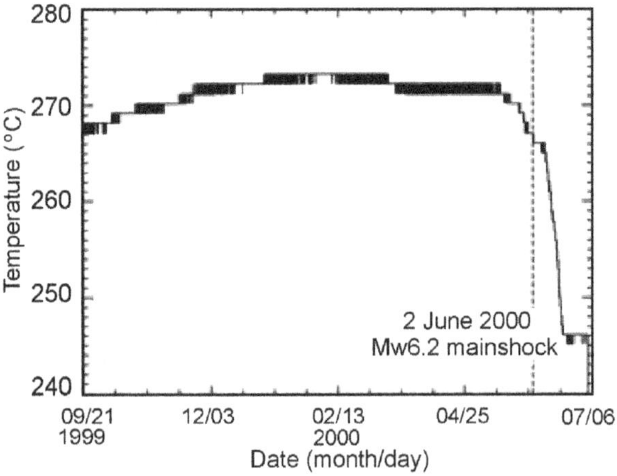

Fig. 6.5 During June 1–7, 2000, an earthquake swarm, with 170 earthquakes and a mainshock of M_W6.2, on the western Blanco Transform Fault. Two temperature probes, located in the hydrothermal vent of an adjacent spreading center, ~39 km away from the earthquake swarm, registered temperature decreases of more than 20°C. One of the records is shown in this figure (From Dziak et al., 2003)

6.2.3.2 Temperature Change in ODP Boreholes on Ridge Flanks

Temperature probes, as well as pore-pressure probes, in ODP boreholes on the eastern flank of the Juan de Fuca Ridge, responded to the June 8, 1999 earthquake swarm on the ridge (Fig. 6.6 from Davis et al., 2001). The fact that both temperature and pore pressure were recorded simultaneously makes this record particularly interesting. Most noteworthy is the observation that while the temperature probes registered changes coinciding not only with the first earthquake, but also with the later earthquakes in the swarm (Fig. 6.6b), the pressure probe registered only a transient change coinciding with the first earthquake (Fig. 6.6a). The coseismic temperature changes were always negative and the amplitudes of the later temperature transients generally reflect the magnitude of the earthquakes, with the greatest change nearly as large as the initial one.

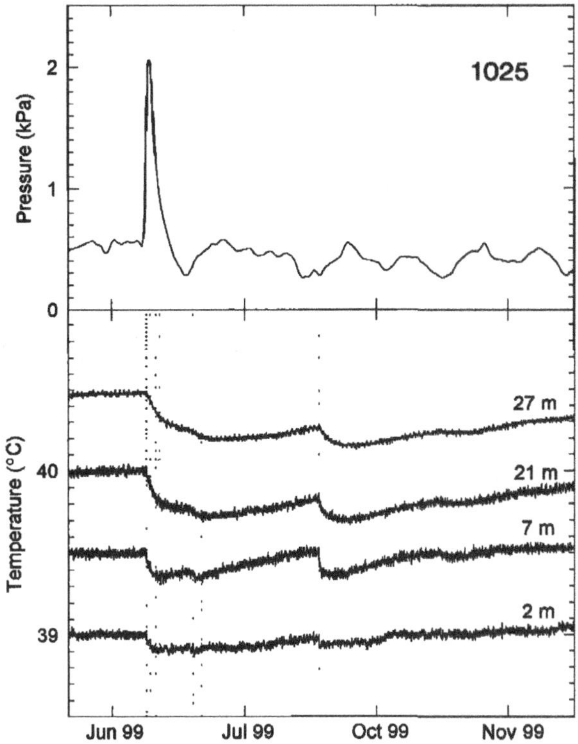

Fig. 6.6 Four-month record of (**a**) pressure and (**b**) temperatures in ODP borehole 1025C on the eastern flank of the Juan de Fuca Ridge, with the times of individual seismic events indicated by *dotted vertical lines*. Depths of the temperature sensors are given relative to the top of basement. The uncased part of the hole extends a total of 47 m below the top of basement. Absolute temperature of the sensor at 21 m is correct; other records are shifted to be offset from this by 0.5 K for plotting convenience. The true gradient in basement is roughly 0.07 K m^{-1} (From Davis et al., 2001)

6.2.4 Mechanisms

6.2.4.1 Hot Springs

As noted earlier, Mogi et al. (1989) hypothesized that the sudden increases of temperature of the hot springs during earthquakes was due to the dislodging of obstacles from the hot spring passageways. This mechanism is similar to the enhanced permeability model proposed in the previous chapters to explain the increase in streamflow (Chap. 4) and sustained change in groundwater level (Chap. 5) after earthquakes. Thus the solid line in Fig. 6.2 that separates earthquakes that cause coseismic changes in spring temperature (solid circles) from those that do not (open circles) provides an independent test of the hypothesis that enhanced permeability may be a common mechanism for several different earthquake-induced hydrological changes. We first express the line in Fig. 6.2 by the following relation:

$$\log r = 0.46M - 0.7 \tag{6.1}$$

and recall the empirical relation among seismic energy density, earthquake magnitude and hypocentral distance, given in Chap. 2 (Sect. 2.4.1):

$$\log r = 0.48M - 0.33 \log e(r) - 1.4. \tag{2.7}$$

Comparing the two relations shows that they have nearly identical slopes on a $\log r$ versus M diagram, and that the two relations nearly coincide when e in relation (2.7) is equal to $\sim 10^{-2}$ J/m^3. In othor words, the solid line that separates the cases with changes in hot spring temperature from those without such changes represents a constant threshold seismic energy of $\sim 10^{-2}$ J/m^3. In the previous chapters, we showed that seismic energy densities greater than 10^{-1} J/m^3 may cause liquefaction and streamflow increase, and that greater than 10^{-3} J/m^3 may cause sustained changes in the groundwater level. The observation of Mogi et al. (1989) provides an independent verification of the general validity of seismic energy density as a metric to relate and delineate various hydrologic responses to earthquakes.

6.2.4.2 Wells

What is the mechanism that causes well water temperature to always drop during the passage of the seismic waves from distant earthquakes (Fig. 6.3)? We can rule out static strain as a possible mechanism because the elastic strain is extremely small due to distant earthquakes (Chap. 1). Furthermore, the observed temperature changes are always negative regardless of earthquake mechanism and the orientation of causative faults, in contrast to what would be expected if static strain were the responsible mechanism. Advection (Appendix C) is an effective process for heat transfer, but since the average velocity and displacement during the water-level oscillations are zero (Fig. 6.3), there should be no net change in temperature due to advection. Shi et al. (2007) suggested that the observed decrease of well water temperature was due to turbulent mixing of well water with stratified temperatures (Pinson et al., 2007). Using finite element modeling and assuming simple dispersion-depth relations in the well, Shi et al. (2007) simulated the earthquake-induced temperature changes in the Tangshan well and showed that the simulated water temperature always drops during the turbulent mixing of well water if heat exchange at the interface between well water and air is sufficiently efficient such that a constant temperature is maintained at this interface. After the water becomes still, thermal conduction becomes dominant, re-establishing the stratified temperatures. Shi et al. (2007) noted, however, that the temperature in the Tangshan well was measured at a single depth and may not provide adequate constraint on the model; they further suggested that, for better constraint, data from a string of high-resolution temperature probes located at selected depths in the well would be required.

6.2.4.3 Marine Geothermal Systems

Temperature changes at hydrothermal vents after earthquake swarms are often interpreted to be the result of opening of clogged cracks and fractures that enhance permeability and flow between reservoirs of different temperatures (e.g., Johnson et al., 2000; Dziak et al., 2003). The substantial delay between the onset of the temperature response and the earthquake swarms may represent the time required for fluids of different temperatures to pass through the newly opened channels and to warm the pathways through which the fluid flows.

As noted earlier, the observation by Davis et al. (2001) in the ODP boreholes on the eastern flank of the Juan de Fuca Ridge is particularly interesting because both pore pressure and temperature were measured simultaneously in the same boreholes. As shown in Fig. 6.6, while pore pressure responded only to the first earthquake in the swarm, temperature in the same boreholes responded to many later earthquakes, in addition to the first one. In the previous chapter we explained why pore pressure responded only to the first earthquake with the model of enhanced permeability; i.e., after the first earthquake, the fluid passageways are cleared and permeability is enhanced, but the time required for the fluid passage to reseal is much longer than the time interval between successive earthquakes in the swarm; thus the local sources were not able to re-pressurize between the successive earthquakes. The reason why temperature responded not only to the first but also to the later earthquakes may be explained by the model of turbulent mixing of the well water with a stratified temperature, as discussed in the previous section. After the water in the borehole becomes still, thermal conduction within the water column re-establishes a stratified temperature profile in a few hours. Thus repeated changes of temperature may appear during the subsequent earthquakes through turbulent mixing.

6.3 Earthquake-Induced Changes in Water Composition

6.3.1 Observations

Most earthquake-related studies of groundwater composition have focused on searching for precursory changes, and we discuss some of these in more detail in Chap. 9. In a few cases, however, the composition of stream water and groundwater were systematically monitored before and after an earthquake. Such changes can provide useful constraints on models of earthquake-induced groundwater flow. Here we review observations from stream water and then those from wells.

6.3.1.1 Change in Stream Water Composition

The stream water chemistries at two gauging stations in the San Lorenzo drainage basin, central California, were monitored on a biannual basis (Fig. 6.7; Rojstaczer and Wolf, 1992). Following the Loma Prieta earthquake of October 17, 1989, stream

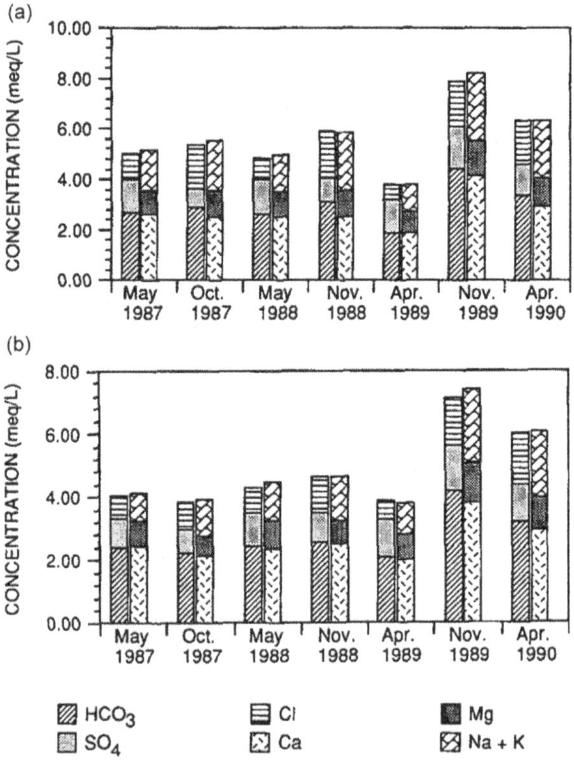

Fig. 6.7 Major ion stream water chemistry as a function of time at two stream gage stations ((**a**) San Lorenzo Park, and (**b**) Big Trees) in the San Lorenzo drainage basin, central California (From Rojstaczer and Wolf, 1992)

water chemistry showed a marked increase in overall ionic strength, but the overall proportions of the major ions were nearly the same as those before the earthquake. The increased ion concentration decreased significantly over a period of several months after the earthquake, together with the decrease in the excess stream discharge. By April 1990, the stream water chemistry had begun to approach the pre-earthquake conditions at both stations. These observations, together with a general cooling of the stream water by several degrees, led Rojstaczer and Wolf (1992, also Rojstaczer et al., 1995) to suggest that the additional stream discharge following the Loma Prieta earthquake was derived from groundwater from the surrounding highlands of the drainage basin.

6.3.1.2 Change in Groundwater Composition

Example from Taiwan

Groundwater samples were collected from a network of monitoring wells on the Choshui River fan before and after the 1999 Chi-Chi earthquake (Fig. 6.8). Because the wells are instrumented with gauges for other purposes, sampling of well water is time consuming and is only done at sparse intervals (Wang et al., 2005). Figure 6.8 shows the measured $\delta^{18}O$ in the groundwater in aquifer II before and after the Chi-Chi earthquake and their differences at several time intervals. Figure 6.8a, b show that the oxygen isotopes in the aquifer shifted significantly towards lighter values after the earthquake. The difference in $\delta^{18}O$ between the post- and pre-earthquake values (Fig. 6.8c–e) decreased with time. In the summer of 2001, nearly two years after the Chi-Chi earthquake, $\delta^{18}O$ in the aquifers returned to their pre-earthquake values (Fig. 6.8e). This recovery time is similar to that found by Claesson et al. (2007) for the earthquake-induced chemical changes in groundwater in Iceland (see next example). It is significantly longer than that of Rojstaczer and Wolf (1992) for the earthquake-induced chemical changes in stream water in central California. Part of the reason for this difference may be the flushing of stream water by surface runoff during the rainy season following the Loma Prieta earthquake.

The mechanism for the observed changes in $\delta^{18}O$ remains unclear. We may rule out the recharge of the confining aquifer by surface runoff because the $\delta^{18}O$ changes were not found in the proximal recharge area of the alluvial fan, but in the remote western coastal region (Fig. 6.8c). It is interesting to note that the pattern of the observed $\delta^{18}O$ change is entirely different from that in the coseismic water-level response in the same aquifer (Fig. 5.4b). Apparently, the mechanism that caused the coseismic water-level changes did not induce a noticeable geochemical changes, and the mechanism that caused the geochemical changes did not induce a significant coseismic water-level change either. Given the proposed mechanism that undrained consolidation may have caused the groundwater level changes in the Choshui River fan during the Chi-Chi earthquake (Chap. 5; Wang et al., 2001), it may be expected that no composition change would be associated with the groundwater-level change. On the other hand, a change in composition must have involved an exchange of water between different sources, given that no

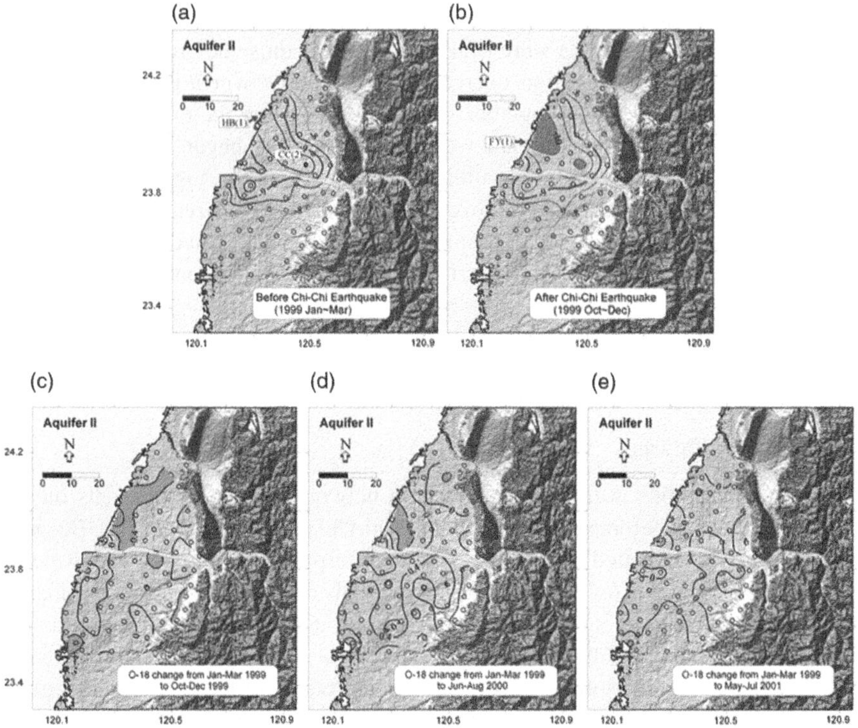

Fig. 6.8 Oxygen isotope contours for Aquifer II in the Choshui alluvial fan from 1999 to 2001. The *open circles* are the sampled wells. (**a**) Absolute $\delta^{18}O$ values in January–March, 1999, before the Chi-Chi earthquake. (**b**) Absolute $\delta^{18}O$ values in October–December, 1999, after the Chi-Chi earthquake; *red* color denotes areas where $\delta^{18}O$ values was below –10‰. (**c**) Difference between (**b**) and (**a**). (**d**) Difference between measurements made in January-to-August, 2000, and (**a**). *Green* color in (**c**) and (**d**) denotes areas where $\delta^{18}O$ was depleted by more than –0.4‰. (**e**) Difference between measurements made in May–July 2001 and (**a**); note that $\delta^{18}O$ has returned to the pre-earthquake values. The scale *bars* are in km (From Wang et al., 2005)

other sources, e.g., water-rock reaction, could have occurred during the short time interval and under the prevailing low temperature. Why isn't there a corresponding change in water level?

This apparent paradox may be explained by considering the Peclet number (Appendix C) for solute transport and that for hydraulic head. For solute transport, $P_e = vL/D$, where v is the linear velocity of the groundwater flow, L is the characteristic distance between different aquifers and D is the coefficient of diffusivity. Thus the discharge of water would cause appreciable change in composition if $v \geq D/L$. For hydraulic head, $P_e = vL/(K/S_s)$; thus the discharge of water would significantly affect the hydraulic head only if $v \geq (K/S_s)/L$ (Phillips, 1991), where K is the hydraulic conductivity and S_s is the specific storage of the aquifer. Since D is of the order of 10^{-11} m^2/s (Ingebritsen et al., 2006), while K/S_s is of the order of 10 m^2/s for the confined aquifers in the Choshui River fan (Tyan et al. 1996), the discharge of water required to cause appreciable change in composition is 12 orders of magnitude smaller than that required to cause appreciable changes in the water level. Thus

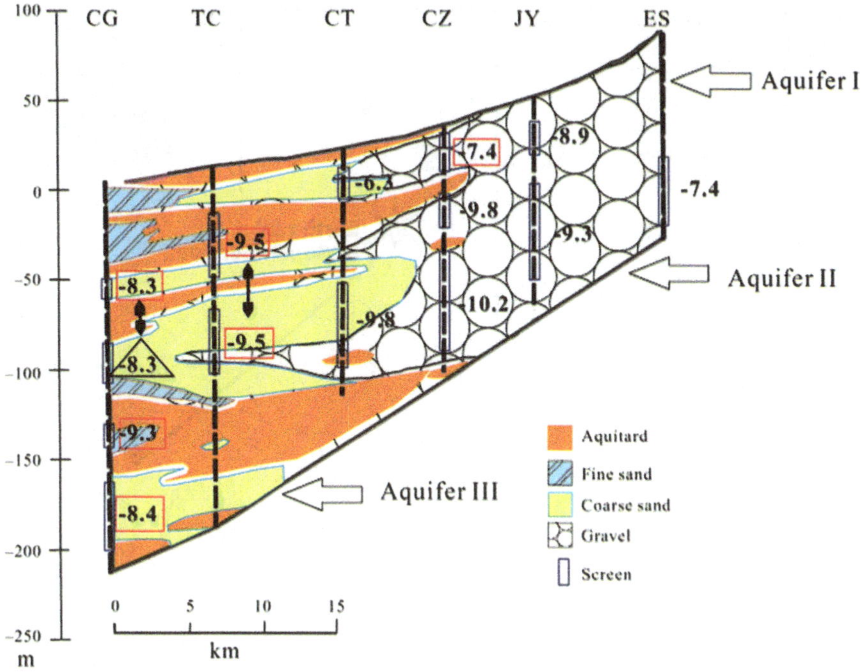

Fig. 6.9 Hydrogeologic cross-section of the Choshui River alluvial fan. The *vertical dashed lines* represent boreholes. The numbers represent the measured $\delta^{18}O$ in units of permil for each sampling site after the Chi-Chi earthquake. The *squares* represent sites that had decreased oxygen isotope values after the earthquake, while the *triangles* represent sites that had increased oxygen isotope values. The double *arrows* indicate aquifers whose $\delta^{18}O$ values converged to the same value after the earthquake (From Wang et al., 2005)

the amount of water exchange to cause the observed composition changes (Fig. 6.8) was far from sufficient to cause an observable change in the groundwater level. A corollary of this scaling analysis is that distinct hydrologic responses occurring simultaneously after an earthquake may show different spatial and temporal patterns depending on different transport processes.

Figure 6.9 shows a vertical cross-section of aquifers beneath the Choshui alluvial fan, together with their $\delta^{18}O$ values following the Chi-Chi earthquake. The $\delta^{18}O$ values in different aquifers were distinct before the earthquake but, after the earthquake, some aquifers showed similar $\delta^{18}O$ values, suggesting that cross-aquifer mixing occurred during or soon after the earthquake, probably due to enhanced vertical permeability through fractures that breached the aquitards between different confined aquifers.

Example from Iceland

Weekly samples were collected from a 1500-m-deep borehole located in a fault zone near Húsavík (Fig. 6.10), Iceland, starting July 2002 (Claesson et al., 2004). A M5.8 earthquake occurred on 16 September 2002 off the coast of Iceland, with epicenter

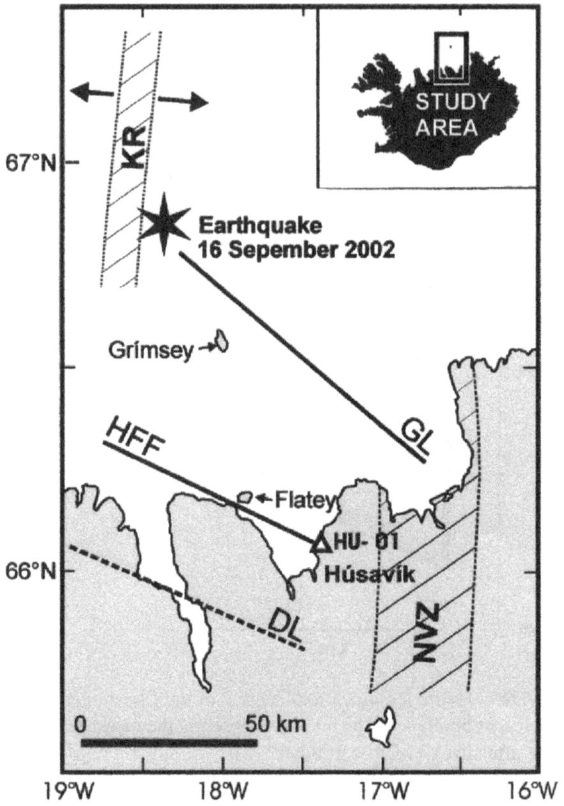

Fig. 6.10 Map showing the geographic locations of the 2002 M 5.8 earthquake off the coast of Iceland, the HU-01 borehole, and major tectonic structures in the region. *Triangle*: HU-01 borehole site; *star*: earthquake epicenter. *NVZ*: Northern Volcanic Zone; *KR*: Kolbeinsey Ridge; *GL*: Grímsey lineament; *HFF*: Húsavík-Flatey fault; *DL*: Dalvík lineament. From Claesson et al. (2004)

$\sim$90 km north of the borehole (Fig. 6.10). The earthquake caused increases by 12–19% in the concentrations of B, Ca, K, Li, Mo, Na, Rb, S, Si, Sr, Cl, and SO_4 in the groundwater and decreases in Na/Ca, $\delta^{18}O$, and δD 2–9 days after the earthquake (Fig. 6.11). The authors suggested that the chemical changes were caused by fluid-source switching to a newly tapped aquifer containing chemically and isotopically distinct water. The connection was enabled by unsealing pre-existing faults.

Claesson et al. (2007) extended their earlier study and found that the chemical changes caused by the M5.8 earthquake recovered gradually over the subsequent

Fig. 6.11 (continued) to Na/Ca data are shown for two models: (1) Na/Ca ratio is unaffected by earthquakes (*blue dotted line*) and (2) 'saw-tooth' variation in Na/Ca ratio, increasing gradually before earthquakes on 16 September 2002 (M_W 5.8), 24 April 2003 (M_W 6.1), and 27 August 2003 (M_W 6.3) and decreasing shortly afterward (red dashed lines). Reproducibility was shown by duplicate analyses. Analytical errors (estimated from replicate analysis of standards) are <2% for Cu, Mn, Zn, Fe, Cr, B, K, Ca, Li, Mo, Na, Rb, S, Si, Sr, Cl, and SO_4, 0.1‰ for $\delta^{18}O$, and 1‰ for δ D (From Claesson et al., 2004)

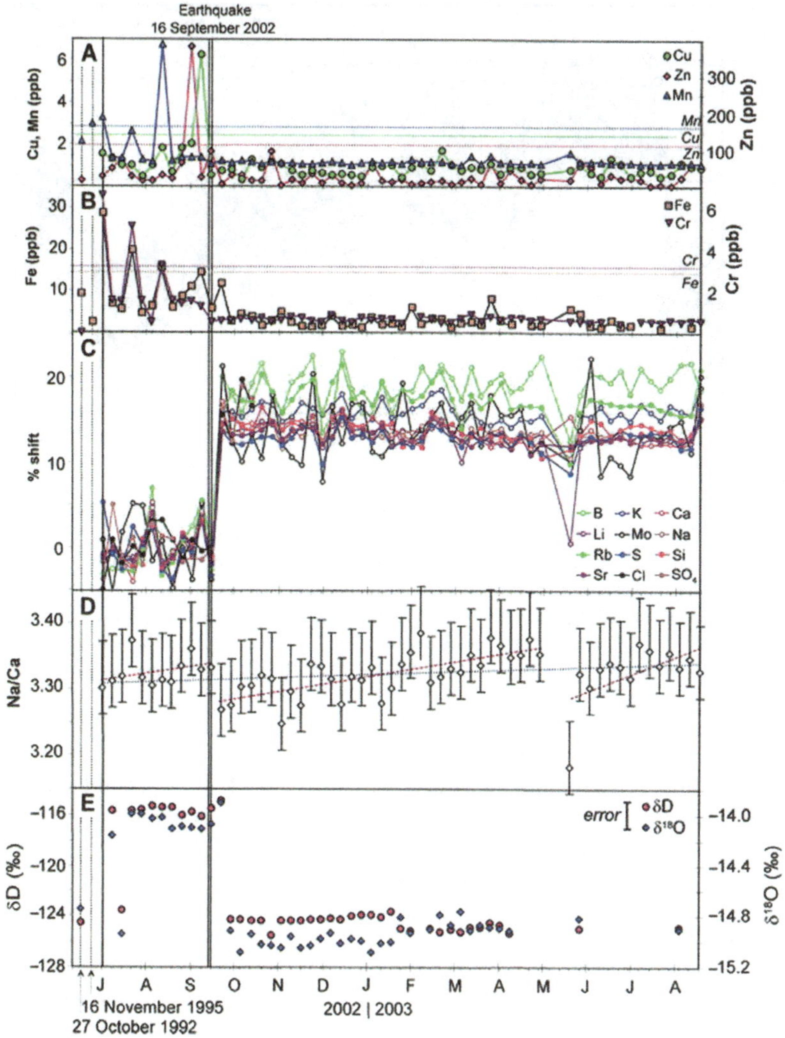

Fig. 6.11 Chemical change before and after an earthquake off the northern coast of Iceland (Fig. 6.10), based on weekly sampling from 3 July 2002 to 28 August 2003; x-axis labels are initial letters of months. **A**: Cu, Mn, and Zn in fluid sampled from 1500-m-deep borehole HU-01. **B**: Fe and Cr. **C**: B, K, Ca, Li, Mo, Na, Rb, S, Si, Sr, Cl, and SO_4, expressed relative to average concentration before earthquake (3 July 2002 to 11 September 2002). **D**: Na/Ca ratio. **E**: $\delta^{18}O$ and δ D. Two-sigma upper limits for 'null hypothesis' that all observed chemical change can be described as normal (Gaussian) variation are shown for Cu, Zn, Mn, Fe, and Cr data sets (*dashed lines*). Probability that Cu, Zn, Mn, Fe, and Cr peaks before earthquake are real anomalies exceeds 99.99%. Note that Cu, Zn, Mn, and Fe anomalies do not arrive at HU-01 at same time. Absolute concentrations of B, K, Ca, Li, Mo, Na, Rb, S, Si, Sr, Cl, and SO4 are available. Linear regressions

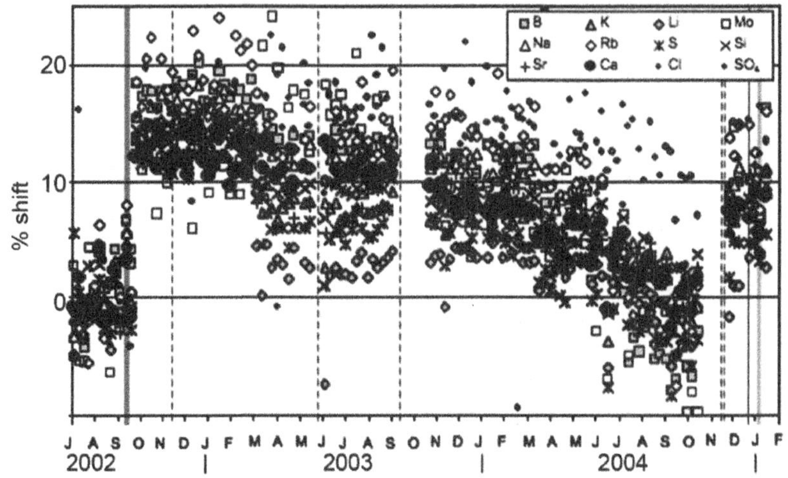

Fig. 6.12 Variation of water chemistry over time for groundwater samples from borehole HU-01. The plot shows percentage shifts in the concentrations of B, K, Li, Mo, Na, Rb, S, Si, Sr, Ca, Cl and SO_4. The timing of $M > 5$, $M > 4$ and $M > 3$ earthquakes are indicated by *red*, *green* and *blue* lines respectively. These lines are solid for earthquakes with strain radii extending beyond the borehole HU-01 and *dashed* for earthquakes with strain radii not reaching HU-01. (Strain radius is defined as the radius of a hypothetical circle about the epicenter of an earthquake, in which the precursory deformations are effectively manifested, Dobrovolsky et al., 1979) (From Claesson et al., 2007)

two years (Fig. 6.12) before the trend was interrupted by a second rapid rise, probably caused by another earthquake.

6.3.2 Mechanisms

Claesson et al. (2007) differentiated between two models for the changes of $\delta^{18}O$ and δ D in groundwater during and after earthquakes. In the first model, accelerated water-rock reactions are caused by an assumed increase in fresh mineral surfaces exposed to groundwater along newly formed cracks and fractures during the earthquake, leading to rapid changes in groundwater composition. In the second model, rapid change in groundwater composition results from fluid-source switching or mixing of groundwater from a newly tapped aquifer containing chemically and isotopically distinct water, probably caused by unsealing of pre-existing faults and breaching of hydrologic barriers. The first model predicts that δ D and $\delta^{18}O$ of groundwater after the earthquake would move away from the Global Meteoric Water Line (GMWL), i.e., along the single-arrowed light-gray curve at the bottom of Fig. 6.13.The second model, on the other hand, predicts that δ D and $\delta^{18}O$ of the groundwater after the earthquake would change in a direction parallel to the GMWL, i.e., along the double arrowed path shown in Fig. 6.13 (Claesson et al., 2007).

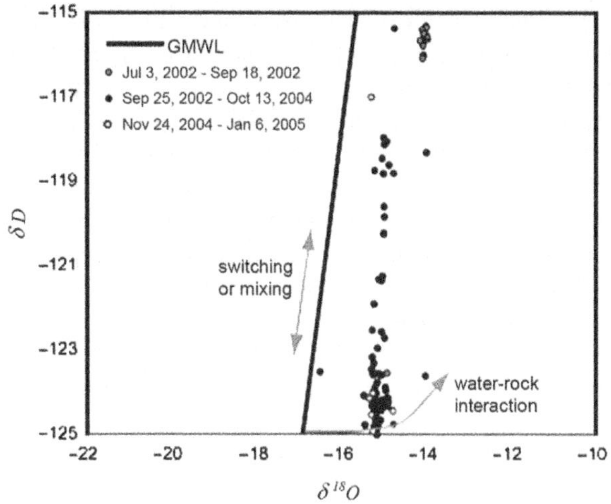

Fig. 6.13 δD versus $\delta^{18}O$ for groundwater samples from borehole HU-01 showing data from 3 July 2002 to 18 September 2002 (*grey circles*), 25 September 2002 to 13 October 2004 (*black circles*) and 24 November 2004 to 6 January 2005 (*white circles*). Both the abrupt hydrogeochemical shift, which occurred with 2–9 days after the M 5.8 earthquake on 16 September 2002, and its recovery during the subsequent two years indicate that switching between or mixing of aquifers is the primary hydrogeochemical control. GMWL is the Global Meteoric Water Line (From Claesson et al., 2007)

The data for three sets of water samples, one collected shortly before and two after the earthquake (Fig. 6.13), show clearly that the changes in $\delta D - \delta^{18}O$ of the groundwater after the earthquake are nearly parallel to the GMWL, consistent with the second model, but inconsistent with the first. Thus the change in the chemical and isotopic compositions of the groundwater after the earthquake was due to source switching and/or mixing with a newly tapped aquifer. This model is consistent with the one suggested earlier to explain the convergence of $\delta^{18}O$ values in different aquifers in the Choshui River fan after the Chi-Chi earthquake (Fig. 6.9; Wang et al., 2005), namely that cross-aquifer mixing occurred during or soon after the earthquake.

6.4 Concluding Remarks

In this chapter we discussed a variety of earthquake-induced changes in water temperature and composition. Even though the observations are in general far less systematic than those for the changes in stream discharge and groundwater level, the available observations provide highly valuable information for constraining models of earthquake-induced groundwater flow. Most changes, such as the temperature change in hot springs and submarine hydrothermal vents and the composition change of groundwater, can be explained by a model of earthquake-enhanced permeability that is due to the dislodging of obstacles from clogged fluid channels, such as pre-existing fractures. Unclogged fractures act to breach hydrologic barriers (such as aquitards) and connect otherwise isolated aquifers or other fluid sources, causing fluid source switching and/or mixing. As discussed in previous chapters, the same model was also used to explain earthquake-induced streamflow increases, pore-pressure re-distribution, and sustained groundwater-level changes.

The sudden decreases of temperature during earthquakes and gradual recovery in some wells and boreholes seem to be caused by an entirely different process, i.e., turbulent mixing in a water column with stratified temperatures. While the validity of the model requires more observational constraint and further quantitative testing, it nonetheless raises a warning flag that such mixing may obscure subtle changes in water temperature brought about by groundwater flow.

The study of earthquake-induced changes in temperature and composition of surface water and groundwater is relatively undeveloped and much is yet to be learned. With advances in technology, measurement of temperature in wells and data transmission have become relatively easy and inexpensive. It may be timely and effective to set up programs for broad monitoring of groundwater temperature in regions with frequent earthquakes.

References

Baker, E.T., C.G. Fox, and J.P. Cowen, 1999, In situ observations of the onset of hydrothermal discharge during the 1998 submarine eruption of Axial volcano, Juan de Fuca Ridge, *Geophys. Res. Lett., 26*, 3445–3448.

Barton, C.A., M.D. Zoback, and D. Moos, 1995, Fluid flow along potentially active faults in crystalline rock, *Geology, 23*, 683–686.

Claesson, L., A. Skelton, C. Graham, C. Dietl, M. Mörth, P. Torssander, and I. Kockum, 2004, Hydrogeochemical changes before and after a major earthquake, *Geology, 32*, 641–644.

Claesson, L., A. Skelton, C. Graham, and C.-M. Mörth, 2007, The timescale and mechanisms of fault sealing and water-rock interaction after an earthquake, *Geofluids, 7*, 427–440.

Davis, E.E., K. Wang, R.E. Thomson, K. Becker, and J.F. Cassidy, 2001, An episode of seafloor spreading and associated plate deformation inferred from crustal fluid pressure transients, *J. Geophys. Res., 106*, 21953–21963.

Department of Monitoring and Prediction of China Earthquake Administration, 2005, *The 2004 M8.7 Sumatra Earthquake, Indonesia, and its Effect on the Chinese Mainland*, Earthquake Publication, Beijing (in Chinese).

Dobrovolsky, I.P., S.I. Zubkov, and V.I. Miachkin, 1979, Estimation of the size of earthquake preparation zones, *Pure Appl. Geophys., 117*, 1025–1044.

Dziak, R.P., W.W. Chadwick, C.G. Fox, and R.W., Embley, 2003, Hydrothermal temperature changes at the southern Juan de Fuca Ridge associated with M-w 6.2 Blanc transform earthquake, *Geology, 31*, 119–22.

Ingebritsen, S.E., W.E. Sanford, and C.E. Neuzil, 2006, *Groundwater in Geologic Processes*, 2nd ed., New York: Cambridge University Press.

Johnson, H.P., M. Hutnak, R.P. Dziak, C.G. Fox, I. Urcuyo, J.P. Cowen, J. Nabelekk, and C. Fisher, 2000, Earthquake-induced changes in a hydrothermal system on the Juan de Fuca mid-ocean ridge, *Nature, 407*, 174–177.

Johnson, H.P., R.P. Dziak, C.R. Fisher, C.G. Fox, and M.J. Pruis, 2001, Impact of earthquakes on hydrothermal systems may be far reaching, *EOS, Trans. Am. Geophys. Union, 82*, 233–236.

Johnson, H.P., J.A. Baross, T.A. Bjorklund, 2006, On sampling the upper crustal reservoir of the NE Pacific Ocean, *Geofluids, 6*, 251–271.

Ma, Z., Z. Fu, Y. Zhang, C. Wang, G. Zhang, and D. Liu, 1990, *Earthquake Prediction: Nine Major Earthquakes in China (1966–1976)*, pp. 332, Beijing: Seismological Press.

Mogi, K., H. Mochizuki, and Y. Kurokawa, 1989, Temperature changes in an artesian spring at Usami in the Izu Peninsula (Japan) and their relation to earthquakes, *Tectonophysics, 159*, 95–108.

Phillips, O.M., 1991, *Flow and Reaction in Permeable Rocks*, Cambridge: Cambridge University Press.

Pinson, F., O. Gregoire, M. Quintard, M. Prat, and O. Simonin, 2007. Modeling of turbulent heat transfer and thermal dispersion for flows in flat plate heat exchangers, *Int. J. Heat Mass Transfer, 50*, 1500–1515.

Rojstaczer, S., and Wolf, S., 1992, Permeability changes associated with large earthquakes: An example from Loma Prieta, California, 10/17/89 earthquake, *Geology, 20*, 211–214.

Rojstaczer, S., S. Wolf, and R. Michel, 1995, Permeability enhancement in the shallow crust as a cause of earthquake-induced hydrological changes, *Nature, 373*, 237–239.

Shi, Y.-L., J.L. Cao, L. Ma, and B.J. Yin, 2007, Tele-seismic coseismic well temperature changes and their interpretation, *Acta Seismo. Sinica, 20*, 280–289.

Sohn, R.A., D.J. Fornari, K.L. Von Damm, J.A. Hildebrand, and S.C. Webb, 1998, Seismic and hydrothermal evidence of a cracking event on the East Pacific Rise near 98509 N, *Nature, 396*, 159–161.

Sohn, R.A., J.A. Hildebrand, and S.C. Webb, 1999, A microearthquake survey of the high-temperature vent fields on the volcanically active East Pacific Rise (98 509 N), *J. Geophys. Res., 104*, 25367–25378.

Tyan, C.L., Y.M. Chang, W.K. Lin, and M.K. Tsai, 1996, The brief introduction to the groundwater hydrology of Choshui River Alluvial fan, in *Conference on Groundwater and Hydrology of the* Choshui *River* Alluvial Fan, *Taiwan*, Water Resources Bureau, 207–221 (in Chinese).

Wang, C.-H., C.-Y. Wang, C.-H. Kuo, and W.-F. Chen, 2005, Some isotopic and hydrological changes associated with the 1999 Chi-Chi earthquake, Taiwan, *The Island Arc, 14*, 37–54.

Wang, C.-Y., L.-H. Cheng, C.-V. Chin, and S.-B. Yu, 2001, Coseismic hydrologic response of an alluvial fan to the 1999 Chi-Chi earthquake, Taiwan, *Geology, 29*, 831–834.

Chapter 7
Geysers

Contents

7.1 Introduction

Geysers are springs that intermittently erupt mixtures of steam and liquid water. Other gases such as CO_2 may play a role in their eruption. Geysers are rare, with probably less than 1000 worldwide, and this number is decreasing owing to geothermal development of the hydrothermal systems they tap (Bryan, 2005). Their rarity reflects the special conditions needed for their formation: a supply of heat that is large enough to boil water, and a plumbing system that has the right geometry to permit episodic discharge. Figure 7.1 shows pictures of some geysers caught in the act of erupting.

Changes in the behavior of geysers is usually characterized by the interval between eruptions, hereafter simply IBE. Geyser eruptions can be periodic (constant IBE), irregular, have a biomodal distribution of IBE, or exhibit chaotic features.

7.2 Response of Geysers to Earthquakes

Geysers have long been known to be particularly sensitive to earthquakes as manifested by changes in IBE. Examples include geysers in California, USA (Silver and Vallette-Silver, 1992) and Yellowstone National Park, USA (e.g., Marler, 1964; Rinehart and Murphy, 1969; Hutchinson, 1985; Husen et al., 2004).

C.-Y. Wang, M. Manga, *Earthquakes and Water*, Lecture Notes in Earth Sciences 114, DOI 10.1007/978-3-642-00810-8_7,

Fig. 7.1 *Left*: Lone Star geyser, Yellowstone National Park. *Middle*: Old Faithful geyser, Yellowstone National Park. *Right*: Strokkur geyser, Iceland, in Geysir basin – the namesake for all geysers. Yellowstone photos taken and provided by Eric Gaidos

There is no systematic pattern to responses after earthquakes. Among the many Yellowstone geysers that have been documented to respond to earthquakes, IBE decreases at some and increases at others (Husen et al., 2004). Figure 7.2 shows the IBE at one Yellowstone geyser and how it responded to 2002 M7.9 Denali earthquake in Alaska, 3100 km away from the geyser. The IBE in general changes suddenly after the earthquake; sometimes the IBE recovers to its pre-earthquake value over a time period of weeks to months, but in other cases the IBE appears to adjust to a new value. Interestingly, some geysers that responded to large earthquakes in 1959 and 1983 did not respond to the 2002 Denali earthquake (Husen et al., 2004).

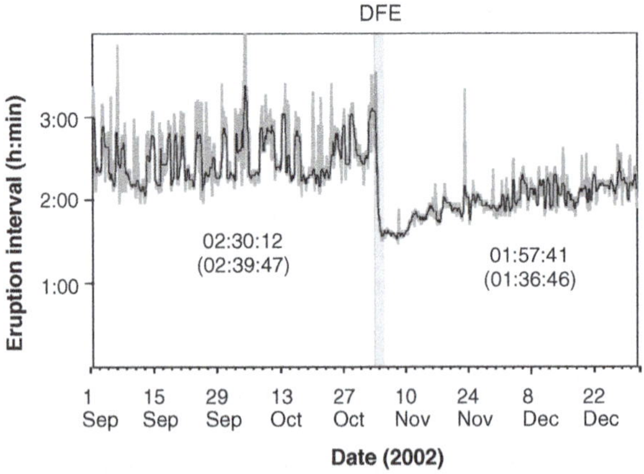

Fig. 7.2 Response of Daisy geyser in Yellowstone to the M 7.9 Denali earthquake located 3100 km from the geyser (Reproduced from Fig. 3 of Husen et al., 2004)

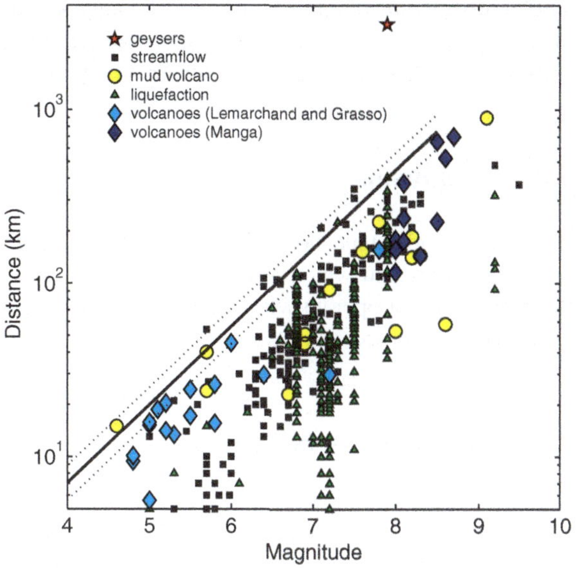

Fig. 7.3 The relationship between earthquake magnitude and distance for a variety of hydrological responses. The Yellowstone geysers, with response to the 2002 M7.9 Denali earthquake shown with the red star, exhibit a sensitivity to earthquakes greater than triggered eruptions of mud volcanoes (*yellow circles*), liquefaction (*black squares*), changes in streamflow (*green triangles*) and volcanoes (*diamonds*). See Appendix E for sources of data

The most remarkable features of the response of geysers to earthquakes is the distance from the epicenter at which they show a sensitivity. They respond to earthquakes that produce static strains $<10^{-7}$ and dynamic strains $<10^{-6}$ (Hutchinson, 1985; Silver and Vallette-Silver, 1992). Figure 7.3 shows that geysers respond to earthquakes at distances far greater than that for the occurrence of liquefaction (Chap. 2) and changes in stream discharge (Chap. 4), but comparable to that for sustained changes in groundwater level (Chap. 5; see also Fig. 10.1).

7.3 Response of Geysers to Other Sources of Stress

The response of geysers to non-seismic strains has been the subject of many studies, and conclusions are not consistent. Some of the inconsistencies may be the result of errors and gaps in eruption catalogs (Nicholl et al., 1994). Earth tides (Rinehart, 1972a, b) and barometric pressure variations (White, 1967) have been reported to influence geyser eruptions in Yellowstone. In a more recent analysis, Rojstaczer et al. (2003) found that Yellowstone geysers are not sensitive to Earth tides or barometric pressure variations – strains typically smaller than 10^{-7}. This is comparable to and larger than the static strains generated by earthquakes that changed eruption intervals. If static earthquake strains caused geysers eruptions to change, we would expect that tides and pressure variations to also cause geysers to change. Only the dynamic strains from earthquakes are much larger than those from tides and pressure variations, so we conclude that the dynamics strains dominate the responses at geysers.

Geysers also respond to hydrological changes. Hurwitz et al. (2008) document clear seasonal variations in IBE and a response to long term trends in precipitation. This indicates that recharge to the geyser plumbing system influences IBE.

7.4 Mechanisms

In order to understand how earthquakes influence geysers it is first necessary to understand how and why geysers erupt. We thus first review published models for the processes that operate within geysers and then identify how earthquakes might influence these processes.

7.4.1 How do Geysers Work?

The evolution of a geyser eruption provides insights and constraints into the processes that lead to their eruption. Geyser eruptions begin with the discharge of water at temperatures below the boiling point; this is followed by a fountain dominated by liquid which progressively becomes more steam-rich before ending with a quiet phase (White, 1967). Bubbles and steam play a central role in transferring heat to warm water in the conduit and in driving the eruption (Kieffer, 1989).

Here we focus on intermittency as this is the property that is documented to change after earthquakes. Two different types of models have been proposed to explain why geysers are intermittent. (1) Ingebritsen and Rojstaczer (1993, 1996) develop a numerical model for groundwater flow and heat transport in an idealized geyser system that consists of a conduit and surrounding matrix, as shown in Fig. 7.4. They show that the observed sequence of events at a geyser can occur periodically for specific combinations of heat flow, conduit and matrix permeabilities, and conduit length. (2) Steinberg et al. (1982 a–c) present a model for geysers in which eruption is driven by the nucleation of steam bubbles in a superheated fluid. The IBE in this case is governed by the time it takes to achieve this degree of superheating.

7.4.2 Mechanisms for Altering Eruptions

Changes in eruption interval can be caused by changes in permeability of the conduit and/or surrounding matrix. As the permeability of the conduit is very high, changes in the matrix that governs conduit recharged are more likely (Ingebritsen and Rojstaczer, 1993). That recharge influences IBE is highlighted by the climate sensitivity of geysers (Hurwitz et al., 2008). Changes in conduit length by reopening blocked and preexisting fractures is an alternative possibility (Ingebritsen et al., 2006). The mechanisms by which the permeability changes or fractures get unblocked remain unclear, but as with hydrological responses reviewed previously, it is likely the dynamic strains that cause such changes. The gradual post-earthquake

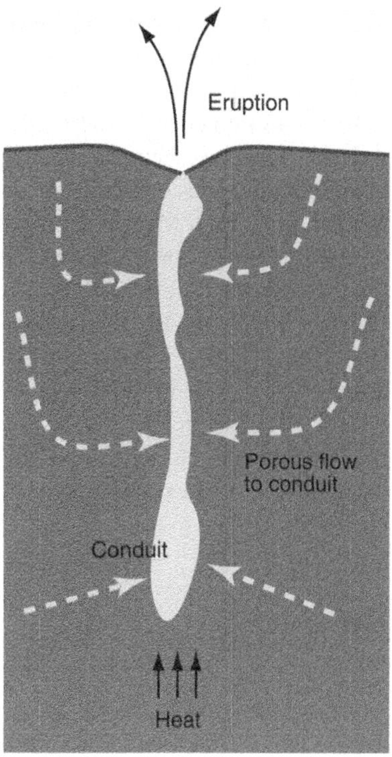

Fig. 7.4 Schematic illustration of a geyser illustrating the plumbing system below the surface (from Manga and Brodsky, 2006). Direct observations of Hutchinson et al. (1997) confirm that the main vent of geyser consists of a complex network of conduits with multiple constrictions

changes in IBE can be explained by gradual fracture sealing and reduction of matrix permeability as has been documented at Yellowstone (e.g., Dobson et al., 2003).

Steinberg et al. (1982c) create a laboratory model of a geyser in which nucleation of bubbles in a superheated fluid drives periodic eruptions. They also show that vibrations can trigger eruptions, presumably by lowering the degree of superheating needed to initiate an eruption. This mechanism does not obviously explain why IBE sometimes increases, or why changes are sustained over multiple eruptions.

7.5 Concluding Remarks

Of the hydrological responses reviewed thus far, geysers stand out because of their extreme sensitivity to seismic waves. The property of geysers used to document these changes is the IBE. With this one measure alone, it is challenging to distinguish between hypotheses about the origin of changes in IBE.

If eruptions are controlled by properties of the geyser plumbing system, because changes in IBE are sustained over multiple eruptions, permanent changes must occur in this plumbing system. Ingebritsen and Rojstaczer (1996) argue changes most likely occur in the recharge to the geyser conduit which is governed by matrix permeability. If IBE is instead controlled by the ability of bubbles to nucleate in a supersaturated system, then it is possible that the earthquake created lower energy nucleation sites that permit eruption at smaller supersaturations.

It may be possible to distinguish between these two end members with additional measurements. In the first case, increased matrix permeability will lead to faster recharge and hence an increase in the mean discharge, though the magnitude of changes depends on details of the conduit and matrix properties (Ingebritsen and Rojstaczer, 1996). Measuring discharge is not easy. The duration of eruption can be used as proxy for discharge, assuming choked flow conditions apply throughout the eruption (Kieffer, 1989). Measurements at Old Faithful geyser, Yellowstone (Kieffer, 1989), the Calistoga geyser, California (Steinberg, 1999) and lab models (Steinberg, 1999) are consistent with IBE scaling with the duration of the previous eruption. If IBE is dominated by nucleation, then the mean discharge will be unaffected – changes in IBE will be accompanied by equivalent changes in the amount of fluid erupted. Simultaneous measurements of discharge and eruption interval may provide the key information to test models and identify the origin of seismic responses.

References

Bryan, T.S. (2005) *Geysers: What they are and How they Work*, 2nd ed., 69 pp, Missoula, MT: Mountain Press Publishing.

Dobson, P.F., T.J. Kneafsey, J. Hulen, and A. Simmons (2003) Porosity, permeability, and fluid flow in the Yellowstone geothermal system, Wyoming, *J. Volcanol. Geotherm. Res., 123*, 313–324.

Hurwitz, S., A. Kumar, R. Taylor, and H. Heasler (2008) Climate-induced variations of geyser periodicity in Yellowstone National Park, USA, *Geology, 36*, 451–454.

Husen, S., R. Taylor, R.B. Smith, and H. Heasler (2004) Changes in geyser eruption behavior and remotely triggered seismicity in Yellowstone National Park induced by the 2002 M 7.9 Denali fault earthquake. *Geology, 32*, 537–540.

Hutchinson, R.A. (1985) Hydrothermal changes in the upper Geyser Basin, Yellowstone National Park, after the 1983 Borah Peak, Idaho, earthquake, in R.S. Stein, and R.C. Bucknam, etc., USGS Open File Report 85-0290A, 612–624.

Hutchinson, R.A., J.A. Westphal, and S.W. Kieffer (1997) In situ observations of old faithful geyser, *Geology, 25*, 875–878.

Ingebritsen, S.E. and S. Rojstaczer (1993) Controls of geyser periodicity, *Science, 262*, 889–892.

Ingebritsen, S.E. and S. Rojstaczer (1996) Geyser periodicity and the response of geysers to deformation, *J. Geophys. Res., 101*, 21891–21905.

Ingebritsen, S.E., W. Sanford, and C. Neuzil (2006) *Groundwater in Geologic Processes*, 2nd ed., 341 pp, New York: Cambridge University Press.

Kieffer, S.W. (1989) Geologic nozzles, *Rev. Geophys., 27*, 3–38.

Manga, M. and E.E. Brodsky (2006) Seismic triggering of eruptions in the far-field: Volcanoes and geysers, *Ann. Rev. Earth Planet. Sci., 34*, 263–291.

Marler, G.D. (1964) Effects of the Hebgen Lake earthquake of August 27, 1959, on the hot spring of the Firehole Geysers basins, Yellowstone National Park, U.S. Geol. Surv. Prof. Paper 435-Q, 185–197.

Nicholl, M.W., S.W. Wheatcraft, and S.W. Tyler (1994) Is old faithful a strange attractor? *J. Geophys. Res., 99*, 4495–4503.

Rinehart, J.S. (1972a) Fluctuations in geyser activity caused by variations in Earth tidal forces, barometric pressure, and tectonic stress, *J. Geophys. Res., 77*, 342–350.

Rinehart, J.S. (1972b) 18.6 year tide regulates geyser activity, *Science, 177*, 346–347.

Rinehart, J.S. and A. Murphy (1969) Observations on pre- and post-earthquake performance of old faithful geyser, *J. Geophys. Res., 74*, 574–575.

Rojstaczer, S., D.L. Galloway, S.E. Ingebritsen, and D.M. Rubin (2003) Variability in geyser eruptive timing and its causes: Yellowstone National Park, *Geophys. Res. Lett., 30*, 1953, doi:10.1029/2003GL-17853.

Silver, P.G. and N.J. Vallette-Silver (1992) Detection of hydrothermal precursors to large northern California earthquakes, *Science, 257*, 1363–1368.

Steinberg, G.S., A.G. Merzhanov, and A.S. Steinberg (1982a) Geyser process: Theory, modeling and field experiment, Part 2. A laboratory models of a geyser, *Mod. Geol., 8*, 61–74.

Steinberg, G.S., A.G. Merzhanov, and A.S. Steinberg (1982b) Geyser process: Theory, modeling and field experiment, Part 3. On metastability of water in geysers, *Mod. Geol., 8*, 75–78.

Steinberg, G.S., A.G. Merzhanov, and A.S. Steinberg (1982c) Geyser process: Theory, modeling and field experiment, Part 4. On seismic influence on geyser regime, *Mod. Geol., 8*, 79–86.

Steinberg, A.S. (1999) An experimental study of geyser eruption periodicity, *Doklady Phys., 44*, 305–308.

White, D.E. (1967) Some principles of geyser activity, mainly from Steamboat Springs, Nevada, *Am. J. Sci., 265*, 641–684.

Chapter 8
Earthquakes Influenced by Water

Contents

8.1 Introduction

Thus far, we have focused on the effects of earthquakes on hydrological processes. This is not a one-way relationship. Changes in pore pressure can also induce earthquakes. In this chapter we thus discuss several ways in which hydrology influences seismicity.

While the term and 'induced' and 'triggered' are often used interchangeably, we endeavor to refer to 'triggered earthquakes' when the hydrologic process adds a small contribution to the stress that causes the earthquake. For 'induced earthquakes', the hydrologic change plays a dominant role. In practice, the distinction can be difficult to make because the mechanisms through which water influences seismicity are not known or straightforward to quantify.

8.2 Fluids and Rock Failure

Earthquakes occur when the stresses acting on rock exceed some critical value so that failure occurs. Fluids influence this failure by changing stress. Consider a fault oriented with angle θ with respect to the largest principle stress, σ_1, as shown in Fig. 8.1. Motion will occur when the shear stress τ on the fault equals or exceeds

C.-Y. Wang, M. Manga, *Earthquakes and Water*, Lecture Notes in Earth Sciences 114, DOI 10.1007/978-3-642-00810-8_8,

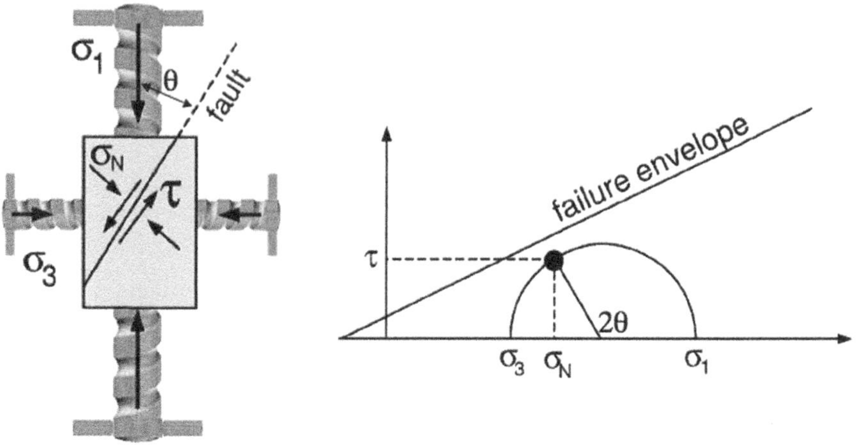

Fig. 8.1 *Left*: Definition of normal and shear stresses on a failure surface oriented at angle θ with respect to the largest principle stress σ_1. *Right*: Graphical representation of the magnitude of normal (horizontal axis) and shear stresses (vertical axis) as a function of the angle θ compared with the magnitude of shear stresses needed for failure (the failure envelope)

the sum of the frictional strength, given by the product of the friction coefficient μ and the normal stress σ_n, and the cohesive strength:

$$\tau = c + \mu\sigma_n \tag{8.1}$$

where c denotes the cohesive strength. Equation (8.1) is usually called Coulomb's failure law (which dates to Coulomb, 1776) or the Mohr-Coulomb failure criterion.

The effect of fluid pressure on failure can be readily understood by realizing that the pressure in the fluid p will support some of the stress. We can thus use the effective stress concept (Terzaghi, 1925) introduced in Chapter 2, to characterize the stresses that influence failure. Because the fluid pressure acts uniformly in all directions, each component in the stress tensor and hence each principle stress will be influenced to the same extent by the fluid pressure (Appendix D),

$$\sigma' = \sigma - \alpha p \tag{8.2}$$

The factor α, called the Biot-Willis coefficient, measures how easily pressure in the fluid is transmitted to the matrix. α is close to 1 for unconsolidated materials, but can be less than 0.5 for rock (Wang, 2000). See Appendix D for more details.

Figure 8.2 shows three ways we can induce failure. In the first two cases, the deviatoric stresses increase, increasing the largest stress, or decreasing the smallest stress, until the Mohr circle intersects the failure envelope. In the third case, or particular interest for the present discussion, an increase in fluid pressure moves the Mohr circle to the left – this is equivalent to maintaining the same shear stresses and reducing normal stresses.

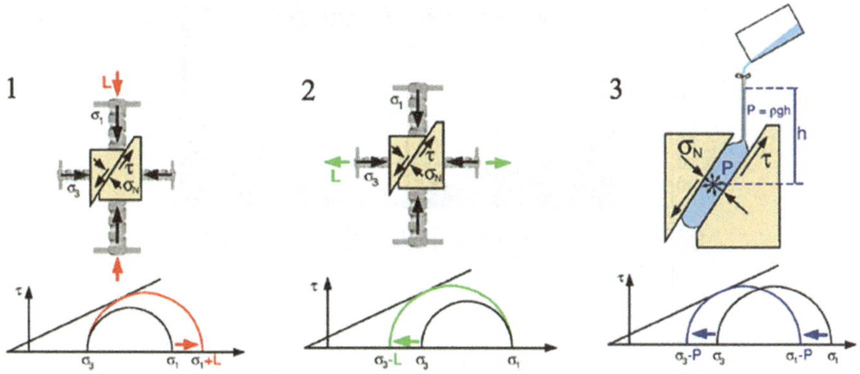

Fig. 8.2 Three ways to induce failure by causing the Mohr circle to intersect the failure envelope: *left*: increasing the largest principle stress σ_1, middle) decreasing the smallest principle stress σ_3, and *right*: increasing fluid press p which reduced the effective stress (Figure 10 from Saar and Manga, 2003)

This Mohr-Coulomb law and the concept of effective stress do not capture the effects of viscous deformation or dilatation that depends on deformation rate and history (such as rate-and-state friction) – processes that have a non-trivial effect on earthquake nucleation and rupture. It is nevertheless useful for illustrating how and why fluids can have a significant influence on earthquakes.

8.3 Earthquakes Induced by Fluid Injection and Extraction

There are both natural and engineering processes that can raise pore pressures and hence influence seismicity. Here we focus on the engineered examples because the sources of fluids are constrained in both space and time. For natural processes, changes in fluid pressure are not in general known, and induced earthquakes, if they can be identified, may provide constraints on these hydrological processes.

A classic and well-studied example of human-induced earthquakes caused by an increase of pore pressure occurred at the Rocky Mountain Arsenal, Colorado, USA. Here, a M 5.5 earthquake occurred, apparently in response to fluid injection at a depth of 3.6 km (Evans, 1966). Continued and controlled monitoring allowed a relationship between injection and seismicity to be established: Figure 8.3 from Healy et al. (1968) shows the history of fluid injected and occurrence of earthquakes. Here, seismicity persisted after injection ended, reflecting the continued diffusion of high pore pressures away from the injection site (Healy et al., 1968; Hsieh and Bredehoeft, 1981).

Earthquakes attributed to fluid injection have occurred along the shore of Lake Erie in Ohio (Seeber et al., 2004), and in the 12 km deep KTB borehole in Germany (Zoback and Harjes, 1997; Bornhoff et al., 2004). In the latter example, Shapiro et al. (2006) document a migration of seismicity that reflects the diffusion of pore

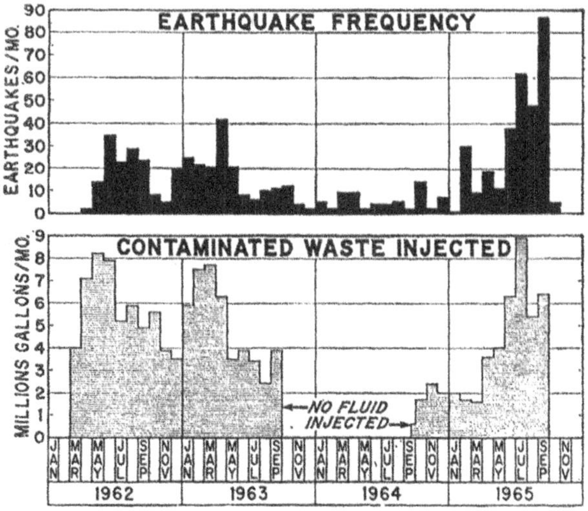

Fig. 8.3 Number of earthquakes recorded at the Rocky Mountain Arsenal waste injection site (*top*) and volume of fluid injected (*bottom*) (From Healy et al., 1968)

pressure along a fault zone. Controlled fluid injection and extraction allow Shapiro et al. (2006) to confirm that seismicity requires increased pore pressure. Talwani (1997) and Tadokoro et al. (2000) identify a similar migration of seismicity along a fault zone following injection. The migration rate of seismicity provides constraints on fault zone hydraulic diffusivity (hence permeability), and when combined with the known pressure at the injection sites, the state of stress on the fault.

Another example of rock failure caused by high pore pressure is hydrofracturing. Here, pore pressure is increased to the point that tensile failure occurs. Hydrofracture is induced intentionally to increase the permeability of rocks in geothermal fields (Majer et al., 2007) or in oil and gas bearing units to enhance recovery.

The concept of effective stress makes it straightforward to understand how injection (i.e. pressure increases) can induce earthquakes. The opposite case, fluid extraction, is also known to induce earthquakes, yet pore pressure reduction acts to stabilize faults. The best documented examples are associated with the extraction of oil and gas (e.g., Segall et al., 1994; Gomberg and Wolf, 1999; Zoback and Zinke, 2002). Segall (1989) shows how poroelastic deformation will increase the magnitude of deviatoric stresses away from the region from which fluid is extracted and where there are no changes in pore-fluid content.

8.4 Reservoir-Induced Seismicity

The filling of surface reservoirs with water also appears to cause earthquakes in some cases. This so-called 'reservoir-induced' seismicity has been documented ever since large reservoirs were constructed (Carder, 1945) and is found on all continents on which reservoirs have been constructed. This topic is reviewed in more detail in a number of books (Gupta, 1992) and review papers (Simpson, 1986; Gupta, 2002).

Earthquakes associated with reservoirs are not confined only to tectonically active regions, hence the reason it is usually called ‘induced’ seismicity. Gupta (2002) notes that the stresses caused by reservoir loading are of order 0.1 MPa, much smaller that earthquake stress drops and hence that these events are better classified as ‘triggered’. Regardless, earthquakes near reservoirs appear to be ubiquitous. Seismicity associated with reservoirs has been documented at passive margins in the United States and South America and within stable cratons in Canada and Africa. As of 2002, 14 different reservoirs experienced earthquakes with magnitude >5 (Gupta, 2002), large enough to cause damage. The Kariba dam between Zambia and Zimbabwe induced a M 6.2 event 5 years after impounding began (Gough and Gough, 1970). Figure 8.4 shows a correlation between water-level increases during the impoundment of the Koyna reservoir, Western India, and the occurrence of earthquakes with $M > 4$, with the largest being a M6.3 event (Gupta and Rastogi, 1976). More recently, water impoundment in a reservoir was implicated as a trigger for the 2008 M 7.9 Wenchuan earthquake in China (see Kerr and Stone, 2009), though this claim was rebuked by Chen (2009). Ge et al. (2009) modeled the effect of reservoir impoundment on pore pressure and the Coulomb failure stress in the

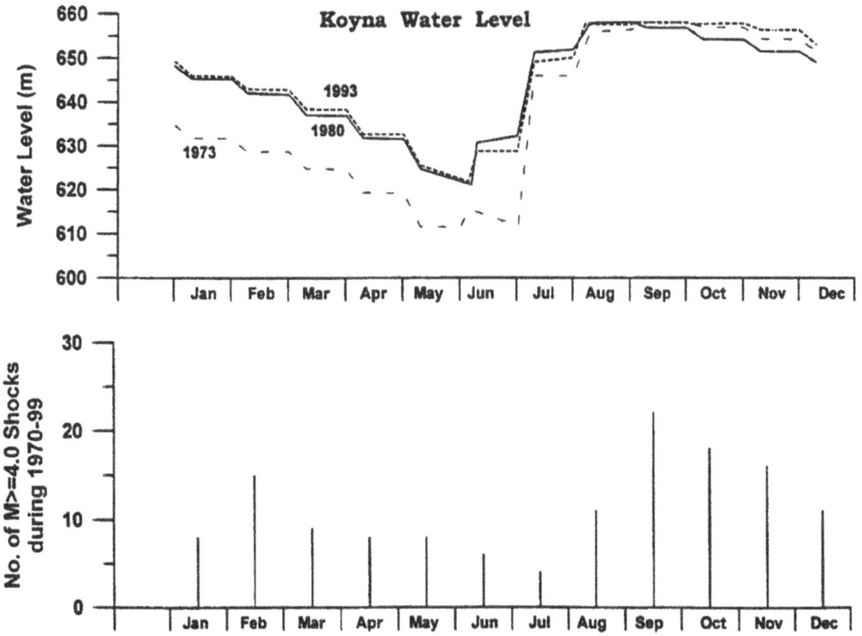

Fig. 8.4 Reservoir-induced seismicity over the period 1970–1999 near the Koyna reservoir, India. *Top*: water level in the reservoir for three representative years. *Bottom*: histograms showing the number of $M > 4$ earthquakes that occur in each month (From Fig. 14 of Gupta, 2002)

surrounding area. Their result shows that the Coulomb failure stress in the hypocentral region of the Wenchuan earthquake increased by 0.01–0.05 MPa within 3 years of the impoundment, and thus promoted the instability of the earthquake fault if an increase of 0.01 MPa in the Coulomb failure stress is significant in a critically stressed region (King et al., 1994; Freed, 2005).

Figure 8.2 may be used to illustrate all the suggested means through which the filling of a reservoir can induce earthquakes (Simpson et al., 1988). The weight of the water can increase both elastic stresses and pore fluid pressure in response to the change in elastic stress – a poroelastic response (e.g., Bell and Nur, 1978). In this case, the orientation of faults and the background stress field will determine where and how faults get reactivated (e.g., Roeloffs, 1988). Groundwater pore pressure should also rise as water seeps out of the reservoir and fluids migrate. The distinctive signature of this second case would be a migration in space of the earthquakes away from the reservoir. This type of migration has been documented at some reservoirs (e.g., Talwani and Acree, 1985).

8.5 Natural Hydrological Triggering of Earthquakes

With insights gained from the engineered occurrences of hydrologically mediated earthquakes, we turn to possible examples of earthquakes triggered by natural hydrological and hydrogeological processes. Here, establishing a connection is more difficult, but potentially more rewarding as it may provide unique insight into hydrogeological and tectonic processes at space and time scales that are otherwise impossible to probe.

Identification of seasonality in seismicity may be indicative of a hydrological influence on earthquakes (Fig. 8.5). This influence could either be in the form of increased stress from the surface load of water or snow, or by changes in pore

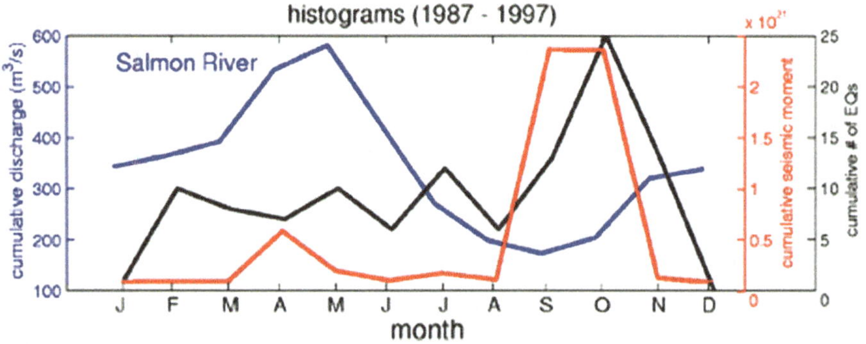

Fig. 8.5 Seasonal variations of seismicity (number of earthquakes in *black*; total moment release in *red*) at Mount Hood, Oregon, USA for the time interval 1987–1997. *Blue* curve shows the mean monthly discharge in a nearby stream and shows the strong seasonality of precipitation. Figure created by Martin Saar

pressure that accompany groundwater recharge. The distinctive signature of the latter, as with reservoir-induced seismicity, is a time lag between the hydrological loading (groundwater recharge) and seismicity.

Seasonal variations of seismicity, while not ubiquitous, have been identified (e.g., Wolf et al., 1997; Bollinger et al., 2007). Heki (2003) identified a seasonal modulation of seismicity in Japan that he attributed to the loading of the surface by snow. Others have attributed seasonal variations of seismicity to groundwater recharge (e.g., Saar and Manga, 2003; Christiansen et al., 2005, 2007); in these cases, the time lag between recharge and seismicity supports an origin from pore pressure changes. A correlation between precipitation and earthquakes (e.g., Roth et al., 1992; Jimenez and Garcia-Fernandez, 2000; Ogasawara et al., 2002; Hainzl et al., 2006; Kraft et al., 2006; Husen et al., 2007) supports the idea that pore pressure changes caused by recharge can influence seismicity (Costain et al., 1987). Figure 8.6 shows the correlation in space and time of earthquakes that appear to be triggered by rainfall in Germany (Hainzl et al., 2006): heavy rain is followed by a downward migration of seismicity, as expected from pore pressure diffusion.

The distinctive signature of earthquakes triggered by pore pressure changes resulting from a localized increase in pore pressure is a spatial migration of locations as time increases. This type of pattern has been recognized for natural earthquake swarms (Noir et al., 1997; Parotidis et al., 2005). A similar idea has been proposed for aftershocks (Nur and Booker, 1972) and characterizes some aftershock sequences (Miller et al., 2004), though rate-and-state friction models can predict migration as well (Toda et al., 2002).

8.6 Earthquake Triggering of Earthquakes via Hydrological Processes

The stresses generated by earthquakes influence the occurrence of additional earthquakes. Many review papers have addressed such connections, including the role of (a) the coseismic static stress changes (e.g., Stein, 1999; King, 2007), (b) dynamic stresses associated with the passage of seismic waves (e.g., Kilb et al., 2000; Hill and Prejean, 2007), and (c) the postseismic relaxation of stresses (Freed, 2005). Here we focus exclusively on mechanisms and processes in which water may play a direct role.

In the near and intermediate field, volumetric strains that accompany earthquakes will change pore pressure. For short times, before significant fluid flow (undrained conditions), the change in pressure Δp scales with the mean stress change σ_{kk} (Appendix D, Eq. D.17)

$$\Delta p = -B\Delta\sigma_{kk}/3 \tag{8.3}$$

where B is the Skempton coefficient. The strain Δe_{ij} can be calculated from the (undrained) constitutive relation (Appendix D, Eq. D.20)

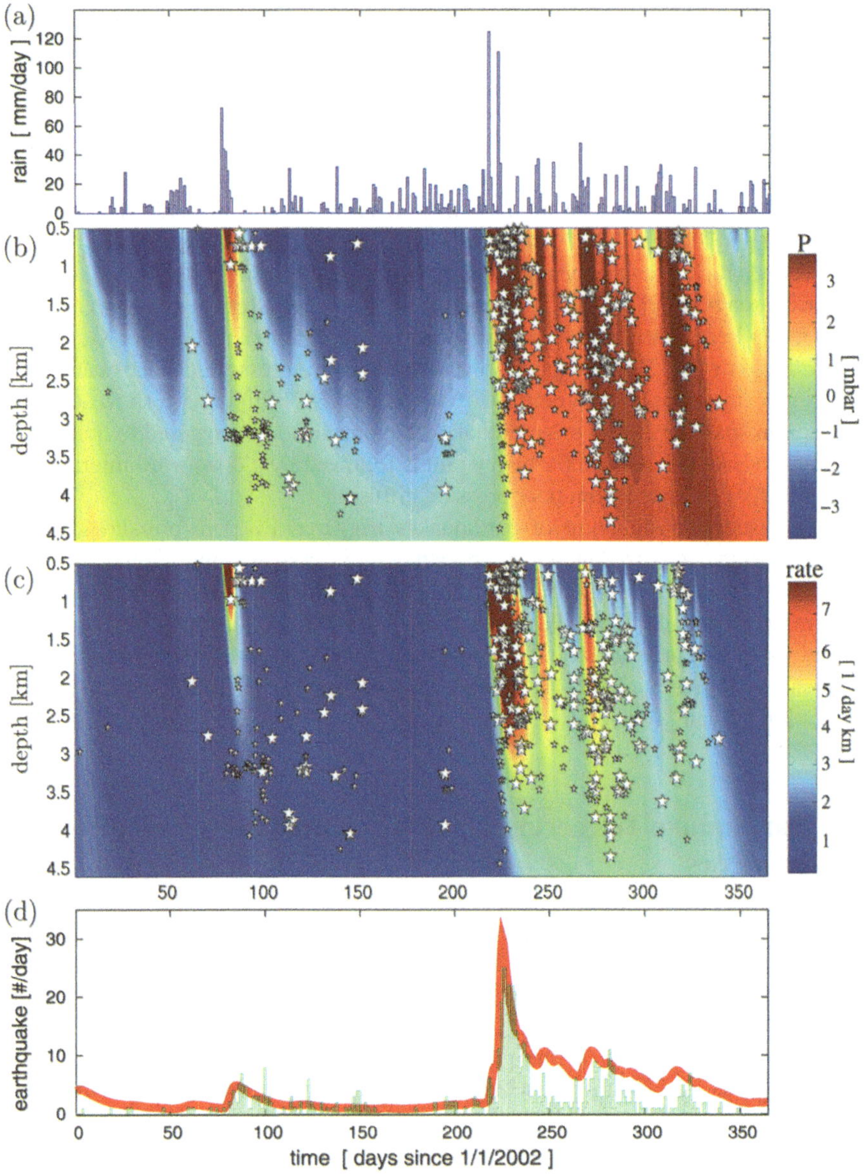

Fig. 8.6 Possible example of seismicity triggered by rainfall-induced changes in pore pressure (from Hainzl et al., 2006). In all panels, **a**–**d**, the horizontal axis is time in days since January 1, 2002. In **b** and **c** the stars show the location of earthquakes; large stars indicate events with location constrained to better than 100 m. **a**) Recorded precipitation. **b**) Predicted distribution of pore pressure assuming one-dimensional diffusion with hydraulic diffusivity (equivalent to $kKB/\eta\alpha$ in Eq. (8.5) of 3.3 m/s^2. **c**) Predicted rate of pore pressure change for the examples shown in **b**. **d**) Daily number of events (*green*) compared with predicted number (*red* curve)

$$2G\Delta e_{ij} = \Delta\sigma_{ij} - \frac{\upsilon_u}{1+\upsilon_u}\Delta\sigma_{kk}\delta_{ij} + \frac{2\,GB}{3}f\delta_{ij} \tag{8.4}$$

where G is the shear modulus and f is the change in fluid content. The undrained Poisson ratio υ_u has values close to 0.33, compared with values typically in the range of 0.15–0.25 under drained conditions (e.g., Wang, 2000) – as the shear modulus should not change under undrained conditions, rock stiffness is greater when undrained than for drained conditions. Cocco and Rice (2002) provide an extension to the case with an anisotropic fault zone imbedded in a surrounding medium with different properties. Cocco and Rice (2002) show that the predicted changes in stress induced by slip are significantly influenced by poroelastic effects.

Postseismic fluid flow will cause a time evolution of stresses and pore pressure. Here the diffusion of pore pressure is governed by the following equation (Appendix D, Eq. D.27):

$$\frac{\alpha}{KB}\left(\frac{B}{3}\frac{\partial\sigma_{kk}}{\partial t} + \frac{\partial p}{\partial t}\right) = \frac{k}{\eta}\nabla^2 p \tag{8.5}$$

where the first terms couples mechanical deformation and fluid flow; k is permeability and η is fluid viscosity. The evolution of flow can be uncoupled from the evolution of stress under a few conditions (e.g., Wang, 2000): the fluid is very compressible (appropriate for gases), and uniaxial strain and constant vertical stress (a reasonable approximation for shallow aquifers). The rate of pore pressure diffusion will scale with the hydraulic diffusivity ($kKB/\eta\alpha$). Pressure changes resulting from postseismic diffusion have been invoked to explain aftershocks (e.g., Nur and Booker, 1972; Bosl and Nur, 2002) and similar seismic sequences (Noir et al., 1997; Antonioli et al., 2005; Miller et al., 2004). The significant non-double-couple component of swarm earthquakes in geothermal areas supports a role of water (Dreger et al., 2000; Foulger et al., 2004). Stress changes will be dominated by changes in p as fluid and fluid pressure are redistributed, but there will be smaller additional changes because the parameters in the constitutive equations will change from their undrained to drained values.

In the far field, fluid flow and poroelastic pressure changes caused by static stress changes are negligible. Triggering of earthquakes in the far-field is thus dominated by dynamic stresses. Examples include seismicity 1250 km away from the M 7.3 Landers earthquake in 1992 (Hill et al., 1993); 1400 km away from the M 8.1 Tokachi-oki earthquake in 2003 (Miyazawa and Mori, 2005); 11,000 km away from the M 8.0 Sumatra earthquake in 2004 (West et al., 2005). Distant triggering is sometimes coincident with the passage of the seismic waves, usually the surface waves that have the greatest amplitudes at these distances. Both Love and Rayleigh waves appear to trigger earthquakes (Velasco et al., 2008). Moreover, triggered events are sometimes even correlated with a particular phase of the waves. West et al. (2005) found that triggered events occur during the maximum horizontal extension associated with the waves. Remote, triggered seismicity need not only be confined to the period of shaking and can sometimes continue for days (e.g., Hill et al., 1993).

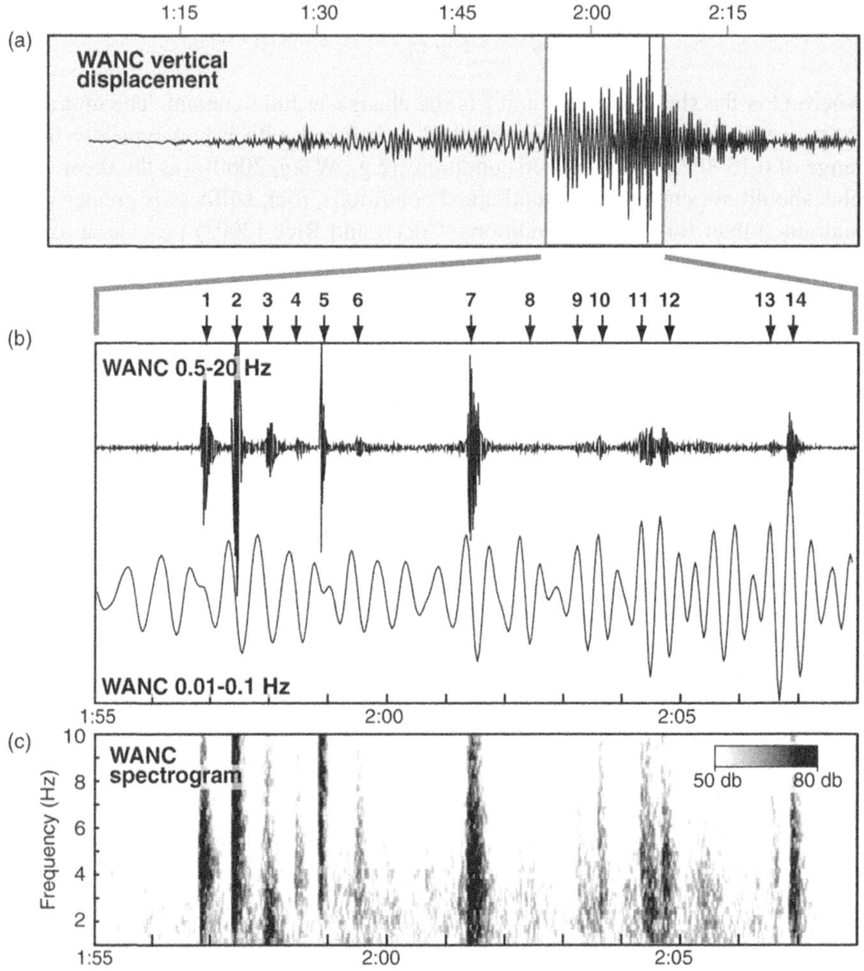

Fig. 8.7 Example of small, local earthquakes triggered near Mount Wrangell, Alaska, by the passage of long period, teleseismic waves from the 26 December, 2004 Sumatra earthquake (from West et al., 2005). **a**) Vertical displacement record. Horizontal scale is one hour. **b**) Expanded view of surface waves shown on top the displacement filtered for 0.5–20 Hz frequencies, and on the *bottom*, 0.01–0.1 Hz. The long period waves are those generated by the Sumatra earthquake; the short period wave packages are created by local, triggered events. Notice that the triggered events always occur during the same phase of the surface wave displacement. **c**) Spectrum of the displacement record as a function of time

Dynamic triggering may be ubiquitous, independent of tectonic environment (Velasco et al., 2008). Nevertheless far-field triggering seems to be better-documented and more abundant in regions with geothermal systems (Spudich et al., 1995; Husen et al. 2004; Manga and Brodsky, 2006) where pore pressures are often (but not always) high – this is suggestive that fluids can play a role in distant triggering. Because dynamic strains are small in the far-field and there is no net strain after the passage of the seismic waves, triggering likely requires a mechanism to translate small and periodic strains into a permanent change in either pore pressure, or permeability (which in turn allows pore pressure to change). The mechanisms discussed in Chaps. 5, 7 and 3 for explaining the response of groundwater level, geysers and volcanoes to earthquakes have also been invoked to explain dynamic triggering of earthquakes: nucleation of new bubbles (Manga and Brodsky, 2006), growth of bubbles (Brodsky et al., 1998), advective overpressure of bubbles shaken loose by seismic waves (Linde et al., 1994), and breaching hydrologic barriers (Brodsky et al., 2003). In contrast, however, West et al. (2005) conclude that their observations (shown in Fig. 8.7) can be simply explained by failure on normal faults caused by the shear stresses generated by the seismic waves. Fluids only need play a role by making the pore pressure high enough that the small dynamic stresses can cause failure.

Identifying whether fluids play any role in causing aftershocks, seismic sequences, or far-field triggering is difficult to confirm observationally because pore pressure measurements are not available. At best, model simulations can be compared with the distribution of earthquakes in space and time, and plausibility can be assessed if the needed parameters are reasonable.

8.7 Concluding Remarks

In previous chapters we saw that earthquakes influence hydrological processes and can change permeability and pore pressure. In this chapter we saw how pore pressure changes can influence seismicity. Figure 8.8 shows some of the various ways in which hydrological processes are coupled to earthquakes.

Rojstaczer et al. (2008) explored some of the consequences of the types of interactions shown in Fig. 8.8 on the evolution of crustal permeability and groundwater flow. They hypothesize that the permeability of the crust adjusts in a time-averaged sense so that it can accommodate recharge by precipitation and fluid released by internal forcing (metamorphism, tectonics). If the pore pressure becomes too large because the permeability is too low, earthquakes occur and will increase permeability. High permeability promotes groundwater flow, mineralization and permeability reduction. As a result, a balance is achieved in which the time-averaged permeability accommodates the transport of fluids provided to the crust. Testing the Rojstaczer et al. (2008) hypothesis is a challenge because of the vast range of space and time scales involved in the processes that influence permeability and groundwater flow. The hypothesis is at least consistent with the mean permeability of the crust (Manning and Ingebritsen, 1999). Other observational evidence supporting this idea, includes mineral deposits that record transient, high permeability flow paths (e.g., Micklethwaite and Cox, 2004) and short-lived high temperatures caused by transient flow in the lower crust (e.g., Camacho et al., 2005).

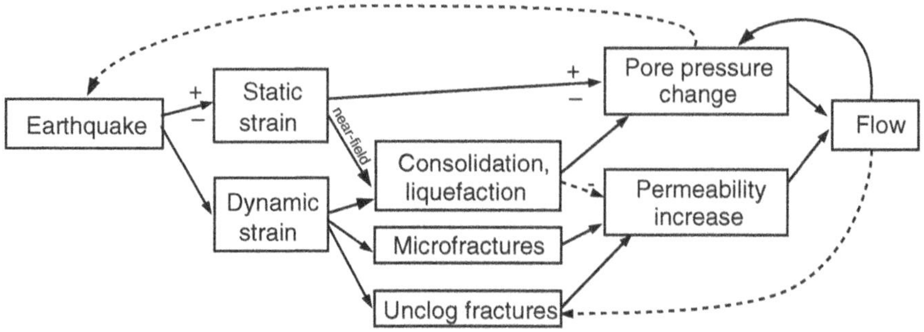

Fig. 8.8 Relationship between earthquakes and hydrology, and the processes through which interactions occur (Figure 17 of Manga and Wang, 2007)

The difficulty in identifying hydrologically induced and triggered earthquakes reminds us that the vast majority of earthquakes are purely tectonic in origin. That is, hydrological changes at Earth's surface do not dominate the occurrence of earthquakes. Triggered earthquakes are nevertheless useful probes of both seismotectonic and hydrologic processes. They can be used to infer hydrogeologic properties and monitor the evolution of subsurface pore pressure, flow, and changes in properties. It is encouraging that more and better instrumentation coupled with a more careful search for such connections are now bearing fruit: hydrologically triggered earthquakes are indeed now routinely identified, and they are proving to be useful to probe and understand subsurface processes.

References

Antonioli, A., D. Piccinini, L. Chiaraluce, and M. Cocco, 2005, Fluid flow and seismicity pattern: Evidence from the 1997 Umbria-Marche (central Italy) seismic sequence, *Geophys. Res. Lett., 32*, L10311, doi:10.1029/2004GL022256.

Bell, M.L. and A. Nur, 1978, Strength changes due to reservoir-induced pore pressure and stresses and application to Lake Oroville, *J. Geophys. Res., 83*, 4469–4485.

Bollinger, L., F. Perrier, J.-P. Avouac, S. Sapkota, U. Gautam, and D.R. Tiwari, 2007, Seasonal modulation of seismicity in the Himalaya of Nepal, *Geophys. Res. Lett., 34*, L08304, doi:10.1029/2006GL029192.

Bornhoff, M., S. Baisch, and H.-P. Harjes, 2004, Fault mechanisms of induced seismicity at the superdeep German Continental Deep Drilling Program (KTB) borehole and their relation to fault structure and stress field. *J. Geophys. Res., 109*, B02309, doi:10.1029/2003JB002528.

Bosl, W.J., and A. Nur, 2002, Aftershocks and pore fluid diffusion following the 1992 Landers earthquake, *J. Geophys. Res., 107*, 2366.

Brodsky, E.E., B. Sturtevant, and H. Kanamori, 1998, Volcanoes, earthquakes and rectified diffusion, *J. Geophys. Res., 103*, 23827–23838.

Brodsky, E.E., E.A. Roeloffs, D. Woodcock, I. Gall, and M. Manga, 2003, A mechanism for sustained ground water pressure changes induced by distant earthquakes. *J. Geophys. Res., 108*, doi:10.1029/2002JB002321.

Camacho, A., J.K.W. Lee, B.J. Hensen, and J. Braun, 2005, Short-lived orogenic cycles and the eclogitization of cold crust by spasmodic hot fluids, *Nature, 435*, 1191–1196.

Carder, D.S., 1945, Seismic investigations in the Boulder Dam area, 1940–1944, and the influence of reservoir loading on local earthquake activity, *Bull. Seism. Soc. Am., 35*, 175–192.

Chen, Y., 2009, Did the reservoir impoundment trigger the Wenchuan earthquake? *Sci. China Ser D-Earth Sci., 52*, 431–433.
Christiansen, L.B., S. Hurwitz, M.O. Saar, S.E. Ingebritsen, and P.A. Hsieh, 2005, Seasonal seismicity at western United States volcanic centers. *Earth Planet. Sci. Lett., 240*, 307–321.
Christiansen, L.B., S. Hurwitz, and S.E. Ingebritsen, 2007, Annual modulation of seismicity along the San Andreas Fault near Parkfield, CA, *Geophys. Res. Lett., 34*, L04306, doi:10.1029/2006GL028634.
Cocco, M., and J.R. Rice, 2002, Pore pressure and poroelasticity effects in Coulomb stress analysis of earthquake interactions, *J. Geophys. Res., 107*, 2030, doi:10.1029/2000JB000138.
Costain, J.K., G.A. Bollinger, and J.A Speer, 1987, Hydroseismicity: A hypothesis for the role of water in the generation of intraplate seismicity. *Seismol. Res. Lett., 58*, 41–64.
Coulomb C.A., 1776, Essai sur une application des regles des maximis et minimis a quelques problemes de statique relatifs a l′architecture. Memoires de l′Academie Royale pres Divers Savants, 7.
Dreger, D.S., H. Tkalcic, and M. Johnston, 2000, Dilational processes accompanying earthquakes in the Long Valley Caldera, *Science, 288*, 122–125.
Evans, D.M., 1966, Denver area earthquakes and the Rocky Mountain Arsenal disposal well, *Mt. Geology, 3*, 23–26.
Foulger, G.R., B.R. Julian, D.P. Hill, A.M. Pitt, P.E. Mailin, and E. Shalev, 2004, Non-double-couple microearthquakes at Long Valley caldera, California, provide evidence for hydraulic fracturing, *J. Volcanol. Geotherm. Res., 132*, 45–71.
Freed, A.M., 2005, Earthquake triggering by static, dynamic and postseismic stress transfer, *Ann. Rev. Earth Planet. Sci., 33*, 335–367.
Ge, S., M. Liu, G. Luo, and N. Lu (2009) Did the Zipingu reservoir trigger the 2008 Wenchuan earthquake, *Geophys. Res. Lett., 36*, L20315, doi:10.1029/2009GL040349.
Gomberg, J., and L. Wolf, 1999, A possible cause for an improbable earthquake: The 1997 M_W 4.9 southern Alabama earthquake and hydrocarbon recovery, *Geology, 27*, 367–370.
Gough, D.I., and W.I. Gough, 1970, Load-induced earthquakes at Lake Kariba – II. *Geophys. J. Int., 21*, 79–101.
Gupta, H.K., 1992, *Reservoir-Induced Earthquakes*, 264 pp, New York: Elsevier.
Gupta, H.K., 2002, A review of recent studies of triggered earthquakes by artificial water reservoirs with special emphasis on earthquakes in Koyna, India, *Earth Sci. Rev., 58*, 279–310.
Gupta, H.K., and B.K. Rastogi, 1976, *Dams and Earthquakes*, 229 pp, Amsterdam: Elsevier.
Hainzl, S., T. Kraft, J. Wassermann, H. Igel, and E. Schmedes, 2006, Evidence for rainfall-triggered earthquake activity, *Geophys. Res. Lett., 33*, L19303, doi:10.1029/2006GL027642.
Healy, J.H., W.W. Ruby, D.T. Griggs, and C.B. Raleigh, 1968, The Denver earthquakes, *Science, 161*, 1301–1310.
Heki, K., 2003, Snow load and seasonal variation of earthquake occurrence in Japan, *Earth Planet. Sci. Lett., 207*, 159–164.
Hill, D.P., and S.G. Prejean, 2007, Dynamic triggering. In: *Treatise of Geophysics, 4*, 257–291.
Hill, D.P., P.A. Reasonberg, A. Michael, W.J. Arabez, G. Beroza, D. Brumbaugh, J.N. Brune, R. Castro, S. Davis, D. dePolo, et al., 1993, Seismicity remotely triggered by the magnitude 7.3 Landers, California, earthquake, *Science, 260*, 1617–1623.
Hsieh, P., and J. Bredehoeft, 1981, A reservoir analysis of the Denver earthquakes: A case of induced seismicity, *J. Geophys. Res., 86*, 903–920.
Husen, S., S. Wiemer, and R.B. Smith, 2004, Remotely triggered seismicity in the Yellowstone National Park region by the 2002 M-W 7.9 Denali fault earthquake, Alaska, *Bull. Seismol. Soc. Am., 94*, 317–331.
Husen, S., C. Bachmann, and D. Diardini, 2007, Locally triggered seismicity in the central Swiss Alps following the large rainfall event of August 2005, *Geophys. J. Int., 171*, 1126–1134.
Jimenez, M.J., and M. Garcia-Fernandez, 2000, Occurrence of shallow earthquakes following periods of intense rainfall in Tenerife, Canary Islands, *J. Volcanol. Geother. Res., 103*, 463–468.
Kerr, R.A., and R. Stone, 2009, A human trigger for the Great Quake of Sichuan? *Science, 323*, 322.

Kilb D., J. Gomberg, and P. Bodin, 2000, Triggering of earthquake aftershocks by dynamic stresses, *Nature, 408*, 570–574.

King, G.C.P., R.S. Stein, and J. Lin, 1994, Static stress changes and the triggering of earthquakes, *Bull. Seism. Soc. Am., 84*, 935–953.

King, G.C.P., 2007, Fault interaction, earthquake stress changes, and the evolution of seismicity. In: *Treatise of Geophysics, 4*, 225–255.

Kraft, T., J. Wassermann, E. Schmedes, and H. Igel, 2006, Meteorological triggering of earthquake swarms at Mt. Hochstaufen, SE-Germany, *Tectonophysics, 424*, 245–258.

Linde, A.T., I.S. Sacks, M.J.S. Johnston, D.P. Hill, and R.G. Bilham, 1994, Increased pressure from rising bubbles as a mechanism for remotely triggered seismicity, *Nature, 371*, 408–410.

Majer, E.L., R. Baria, M. Stark, S. Oates, J. Bommer, B. Smith, and H. Asanuma, 2007, Induced seismicity associated with enhanced geothermal systems, *Geothermics, 36*, 185–222.

Manga, M., and E. Brodsky (2006) Seismic triggering of eruptions in the far-field: Volcanoes and geysers, *Ann. Rev. Earth Planet. Sci., 34*, 263–291.

Manga, M., and C.-Y. Wang (2007) Earthquake hydrology. In: *Treatise of Geophysics*, G. Schubert editor, Vol. 4, pp. 293–320.

Manning, C.E. and S.E. Ingebritsen, 1999, Permeability of the continental crust: Constraints from heat flow models and metamorphic systems, *Rev. Geophys., 37*, 127–150.

Micklethwaite, S. and S.F. Cox, 2004, Fault-segment rupture, aftershock-zone fluid flow, and mineralization, *Geology, 32*, 813–816.

Miller, S.A., C. Collettini, L. Chiaraluce, M. Cocco, M. Barchi, and B.J.P. Kaus, 2004, Aftershocks driven by a high-pressure CO2 source at depth, *Nature, 427*, 724–727.

Miyazawa, M., and J. Mori (2005) Detection of triggered deep low-frequency events from the 2003 Tokachi-oki earthquake, *Geophys. Res. Lett., 32*, L10307, doi:10.1029/2005GL022539.

Noir, J., E. Jacques, S. Bekri, P.M. Adler, P. Tapponnier, and G.C.P. King, 1997, Fluid flow triggered migration of events in the 1989 Dobi earthquake sequence of Central Afar, *Geophys. Res. Lett., 24*, 2335–2338.

Nur, A., and J.R. Booker, 1972, Aftershocks caused by pore fluid flow? *Science, 175*, 885–887.

Ogasawara, H., Y. Kuwabara, T. Miwa, K. Fujimore, N. Hirano, and M. Koizumi, 2002, Postseismic effects of an M 7.2 earthquake and microseismicity in an abandoned, flooded, deep mine, *PAGEOPH, 159*, 63–90.

Parotidis, M., S.A. Shapiro, and E. Rothert, 2005, Evidence for triggering of the Vogtland swarms 2000 by pore pressure diffusion, *J. Geophys. Res., 110*, B05S10, doi:10.1029/2004JB003267.

Roeloffs, E.A., 1988, Fault stability changes induced beneath a reservoir with cyclic variations in water level, *J. Geophys. Res., 93*, 2107–2124.

Rojstaczer, S.A., S.E. Ingebritsen, and D.O. Hayba, 2008, Permeability of continental crust influenced by internal and external forcing, *Geofluids, 8*, 128–139.

Roth, P., N. Pavoni, and N. Deichmann, 1992, Seismotectonics of the Eastern Swiss Alps and evidence for precipitation-induced variations of seismic activity, *Tectonophysics, 207*, 183–197.

Saar, M.O., and M. Manga, 2003, Seismicity induced by seasonal groundwater recharge at Mt. Hood, Oregon, *Earth Planet. Sci. Lett., 214*, 605–618.

Seeber, L., J.G. Armbruster, and W.-Y. Kim, 2004, A fluid-injection-triggered earthquake sequence in Ashtabula, Ohia: Implications for seismogenesis in stable continental regions, *Seism. Soc. Am., 94*, 76–87.

Segall, P., 1989, Earthquakes triggered by fluid extraction, *Geology, 17*, 942–946.

Segall, P., J.R. Grasso, and A. Mossop, 1994, Poroelastic stressing and induced seismicity near the Lacq gas field, southwestern France, *J. Geophys. Res., 99*, 15423–15438.

Simpson, D.W., 1986, Triggered earthquakes, *Ann. Rev. Earth Planet. Sci., 14*, 21–42.

Simpson, D.W., W.S. Leith, and C.H. Scholz, 1988, Two types of reservoir-induced seismicity. *Bull. Seism. Soc. Am., 78*, 2025–2050.

Shapiro, S.A., J. Kummerow, C. Dinske, G. Asch, E. Rothert, J. Erzinger, H.-J. Kmpl, and R. Kind, 2006, Fluid induced seismicity guided by a continental fault: Injection experiment of 2004/2005 at the German Deep Drilling Site (KTB), *Geophys. Res. Lett., 33*, L01309, doi:10.1029/2005GL024659.

Spudich, P., L.K. Steck, M. Hellweg, J.B. Fletcher, and L.M. Baker, 1995, Transient stresses at Parkfield, California, produced by the M 7.4 Landers earthquake of June 28, 1992 Observations from the USPAR dense seismograph array, *J. Geophys. Res., 100*, 675–690.

Stein, R.S., 1999, The role of stress transfer in earthquake occurrence. *Nature, 402*, 605–609.

Talwani, P., 1997, On the nature of reservoir-induced seismicity, *PAGEOPH, 150*, 473–492.

Talwani, P., and S. Acree, 1985, Pore pressure diffusion and the mechanism of reservoir-induced seismicity, *PAGEOPH, 122*, 947–965.

Tadokoro, K., M. Ando, and K. Nishigami, 2000, Induced earthquakes accompanying the water injection experiment at the Nojima fault zone, Japan: Seismicity and its migration, *J. Geophys. Res., 105*, 6089–6104.

Terzaghi, K., 1925, *Erdbaumechanik auf bodenphysikalischer Grundlage*, Leipzig: Deuticke.

Toda, S., R.S. Stein, and T. Sagiya, 2002, Evidence from the AD 2000 Izu Islands earthquake swarm that stressing rate governs seismicity, *Nature, 419*, 58–61.

Velasco, A.A., S. Hernandez, T. Parsons, and K. Pankow, 2008, Global ubiquity of dynamic earthquake triggering, *Nature Geosci., 1*, 375–379.

Wang, H.F., 2000, *Theory of Linear Poroelasticity*, 287 pp, Princeton, NJ: Princeton University Press.

West, M., J.J. Sanchez, and S.R. McNutt, 2005, Periodically triggered seismicity at Mount Wrangell, Alaska, after the Sumatra earthquake, *Science, 308*, 1144–1146.

Wolf, L.W., C.A. Rowe, and R.B. Horner, 1997, Periodic seismicity near Mt. Ogden on the Alaska-British Columbia border: A case for hydrologically-triggered earthquakes? *Bull. Seismol. Soc. Am., 87*, 1473–1483.

Zoback, M.D., and H. Hanjes, 1997, Injection-induced earthquakes and crustal stress at 9 km depth at the KTB deep drilling site, Germany, *J. Geophys. Res., 102*, 18477–18492.

Zoback, M.D., and J. Zinke, 2002, Production-induced normal faulting in the Valhall and Ekofisk oil fields, *PAGEOPH, 159*, 403–420.

Chapter 9
Hydrologic Precursors

Contents

9.1 Introduction

The ability to recognize precursory signals would clearly be of tremendous value. For this reason the search for precursors to earthquakes has a long history. Despite early optimism (Scholz et al., 1973), the results so far have been disappointing.

There are mechanical reasons for anticipating precursors. Laboratory studies of rock deformation show that beyond the elastic limit, shearing of consolidated rock creates microcracks that open and hence increase the rock volume (e.g., Brace et al., 1966). At still higher deviatoric stresses, microcracks merge and localize to form a shear zone, leading to eventual large-scale rupture (e.g., Lockner and Beeler, 2002). There are several ways in which the mechanical changes leading up to rupture can be manifest in hydrological measurements. First, the increase in surface area

C.-Y. Wang, M. Manga, *Earthquakes and Water*, Lecture Notes in Earth Sciences 114, DOI 10.1007/978-3-642-00810-8_9, © Springer-Verlag Berlin Heidelberg 2010

produced by microcracks can release gases trapped in pores (e.g., radon) or change the ionic concentration and hence electrical conductivity of groundwater. Such possible changes provide the motivation for seeking and interpreting changes in gas concentration, hydrogeochemistry, or electrical conductivity. Second, microcracks can change hydrogeologic properties such as permeability as well as pore pressure. Such changes, in turn, can cause a redistribution of fluids and fluid pressure and hence may be detected from changes in water level in wells or changes in spring and stream discharge.

As reviewed in previous chapters, hydrologic systems can greatly magnify minute tectonic and seismic strains, as recorded by changes in pore pressure and water level in wells. For example, the change in pore pressure p under undrained conditions is given by (Appendix D, Eq. D.17)

$$p = -K_u B \varepsilon_{kk} \tag{9.1}$$

where K_u is the undrained bulk modulus and B Skempton's coefficient. A volumetric strain $\varepsilon_{\mathrm{kk}}$ as small as 10^{-8} can be expected to produce (detectable) water changes of 2 cm for reasonable choices of $B = 1$ and $K_{\mathrm{u}} = 20$ GPa (Wang, 2000). It is in part because of the potential sensitivity of hydrogeological systems that much of the search for precursors has focused on hydrological measurements. In addition, hydrological measurements can be made with relative ease (compared with electromagnetic and seismic surveys) and can be recorded continuously.

The hydrogeochemical basis for searching for precursors is similar. The gas composition of springs, for example, can respond to (small) tidal strains (e.g., Sugisaki, 1981), hence any preseismic strain might be amplified by hydrogeochemical changes. Radon concentration changes are the most commonly reported and discussed hydrogeochemical precursor (e.g., King, 1980; Wakita et al., 1988; Virk and Singh, 1993) and geochemical recorder of small strains (e.g., Trique et al., 1999) – this is not unreasonable given that radon accumulates over time in micropores, and can be released by small structural changes in rocks and pore connectivity. Small strains may also permit mixing between reservoirs by breaching barriers, or may expose fresh mineral surfaces which in turn permit water-rock interaction (e.g., Thomas, 1988). In a manner similar to hydrological recovery after co-seismic hydrological changes (stream flow, water level in wells), water geochemistry also exhibits a postseismic recovery if disturbed by the earthquake (e.g., Claesson et al., 2007).

Because the seismic wave velocities are highly sensitive to the opening and closing of microcracks and to the changes in their degree of saturation (e.g., O'Connell and Budiansky, 1974), seismologists have carried out various experiments to test the microcrack hypothesis and produced a series of controversial results over the past 50 years. The first published works of such tests were carried out by Kondratenko and Nersesov (1962) for earthquakes in the Tadjikistan region and by Semenov (1969) for earthquakes near Garm, both in the former Soviet Union. These reports were initially met with skepticism by seismologists in both Japan and the United States (Bolt and Wang, 1975). Nevertheless, the work was sufficiently suggestive to motivate other seismologists to set out independent experiments to examine the claims. The first U.S. experiments along these lines, using quite small earthquakes in the Adirondacks in New York, also detected reductions in the Vp/Vs ratio in three cases (Aggarwal et al., 1973). After the 1971 San Fernando earthquake (magnitude 6.5), Whitcomb et al. (1973) concluded that there had been a precursory decrease in the Vp/Vs ratio lasting about 30 months and a subsequent return to normal, which was followed quickly by the earthquake. On the other hand, McEvilly and Johnson (1974) used travel times between quarry blasts in central California along the San Andreas fault, with known position and origin time, and the University of California seismic network; their study indicated that the recorded fluctuations in travel times for the years 1961–1973 could be accounted simply by reading errors and changes of shot location in the quarry. They concluded that there were no detectable premonitory travel-time changes prior to 17 earthquakes in the region with magnitudes between 4.5 and 5.4. Later work (Robinson et al., 1974) in the region indicates, however, that positive P residuals were detectable before the 1972 Bear Valley earthquake (magnitude 5.1) when carefully located local earthquakes were used as sources and the wave paths passed close to the focal region. Recently, Niu et al. (2008) conducted an active source cross-well experiment at the San Andreas Fault Observatory at Depth (SAFOD) drill site and studied the shear wave travel time along a fixed pathway for three small earthquakes ($M \leq 3$) over a period of 2 months. They show excursions in the travel time before two of these earthquakes, but no excursion before the third. More experiments, with longer duration and careful planning, are obviously needed to explain these differences.

9.2 What is a Precursor?

We begin by defining a 'precursor' as a change in a measured quantity that occurs prior to an earthquake that does not originate from any process other than those that lead to the earthquake. Reported hydrological examples include changes in water pressure, streamflow, and water geochemistry and turbidity.

A useful precursor is one that also predicts the time, location and size of the forthcoming earthquake. To our knowledge, no paper has claimed to make these three predictions based on reported hydrologic anomalies.

9.3 Identifying Hydrologic Precursors

Definitive and consistent evidence for hydrological and hydrogeochemical precursors has remained elusive (Bakun et al., 2005) to the extent that there is no consensus on the significance and origin of reported precursors. Difficulties include that, until recently, most reported changes were not corrected for the fluctuations in temperature, barometric pressure, earth tides, and other environmental factors, so that some changes taken to be earthquake-related may in fact be 'noise' (e.g., Hartman and Levy, 2005). One common feature of reports is that changes are recorded at some sites but not at other nearby sites (e.g., Biagi et al., 2001). Moreover, instrument failures and personnel/program changes often do not allow persistent and consistent monitoring over long periods of time (King et al., 1994) – a necessary condition for obtaining reliable precursory data. Distinguishing a precursor from a response to a previous earthquake creates additional ambiguity.

Roeloffs (1988) lists the ideal, and arguably necessary, criteria and complementary data for establishing that some hydrologic signal is in fact precursory. We reproduce (sometimes paraphrased or modified slightly) her list below and comment on some of these criteria afterwards. As noted by Roeloffs (1988), poor documentation is the major impediment to using and interpreting water level data.

(1) Depth of well
(2) Rainfall over at least one year
(3) Record of barometric pressure recorded at least once every three hours
(4) Information about pumping and injection at wells in the vicinity
(5) The entire observation record should be presented
(6) Measurement technique (e.g., pressure transducer, float)
(7) Sampling interval; this should be short enough to reliably distinguish between anomalies before and after the earthquake (Sugisaki, 1978)
(8) Response to earth tides
(9) Co-seismic and post seismic response to the earthquake
(10) Earthquake magnitude, azimuth, distance, depth and focal mechanism
(11) Time, location and magnitude of any foreshocks
(12) Raw water level data (unprocessed)
(13) Description of other wells in the area that did not document the anomaly.

Roeloffs (1988) also points out that site geology, in particular the proximity to fault zones, and whether the reservoir is confined, are useful for interpreting any signals. For gas and hydrogeochemical anomalies, multiple measurements of ions and gases are helpful in identifying the origin and reliability of the anomaly (Sugisaki and Suguira, 1985).

The importance of removing signals that arise from tides and barometric pressure variations is highlighted in Fig. 9.1 in which raw water level records are compared with records in which the effects of tides and barometric pressure changes are removed. The coseismic water level response becomes much more clear in the latter. In addition, two pre-seismic anomalous changes become apparent (discussed in more detail later).

Notwithstanding these difficulties, progress has been made in the past decade. For example, intensive and continued observations of various kinds of precursory hydrological and hydrogeochemical changes have been made in Japan during the past half century (Wakita, 1996) – providing a long time series of observations. Records are now routinely corrected to remove the noise introduced by fluctuations in temperature, barometric pressure, earth tides, and other factors (Igarashi and Wakita, 1995). Tools are readily available to remove the effects of earth tides and barometric pressure variations (e.g., Toll and Rasmussen, 2007). The importance of these corrections should be clear from all the raw records presented in

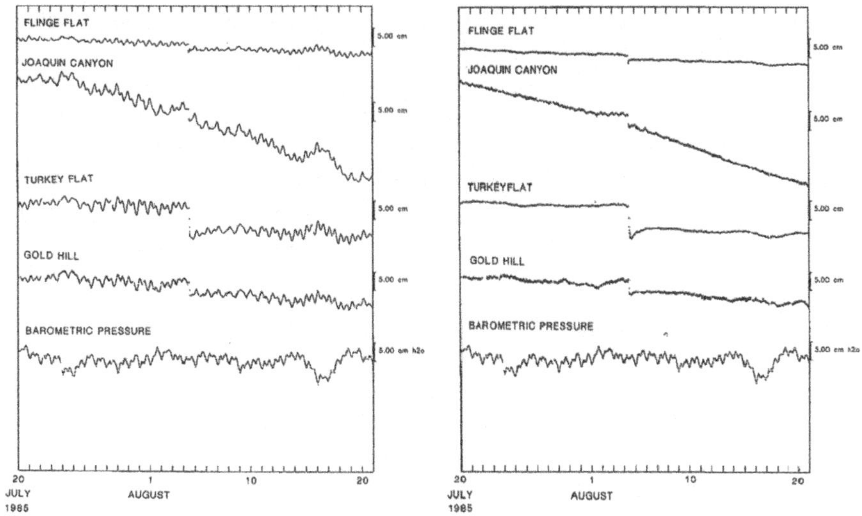

Fig. 9.1 Raw water level measurements (*left*) at 4 wells located near Parkfield, California. The coseismic response to the August 4 1985 M 6.1 Kettleman Hill, California earthquake can be clearly seen in the raw water level records. Earthquake was located 30–40 km from these wells. On the *right* is the same data with the effects of earth tides and barometric pressure removed. The coseismic response remains clear. Now, two proposed precursory signals can be seen, a gradual preseismic increase in Joaquin Canyon and Gold Hill (Figures from Quilty and Roeloffs, 1997)

this chapter. Other signal processing techniques can be helpful. For example, high- and low-pass filtering has been applied to the time series of raw hydrogeochemical data in Kamchatka, Russia, to remove long- and short-period changes unrelated to earthquake processes (Kingsley et al., 2001).

Effort has also been made to address the statistical significance of possible precursors. Statistical, rather than deterministic, procedures have been introduced (Maeda and Yoshida, 1990) to assess the conditional probability of future seismic events. Multi-component, hydrochemistry analysis was applied to groundwater samples in Iceland before and after a major earthquake to enhance the possibility of detecting possible precursors (Claesson et al., 2004). Highlighting the importance of long time series, Claesson et al. (2007) extended the time series of geochemical measurements after this (and subsequent) earthquakes and found that the statistical significance of previously identified anomalies could not be verified.

9.4 Examples

There are hundreds of reports of possible earthquake precursors. Here we review and discuss only selected studies to (1) illustrate the range of types of measurements that have been made, (2) the challenges with identifying precursors, and (3) some of the key questions raised by reported precursor identifications. In all but one of the examples that we discuss, the hydrological changes are identified retrospectively as being premonitory to the earthquake.

Reviews of reported hydrologic precursors include Roeloffs (1988) and Hartmann and Levy (2005); hydrogeochemical precursor reviews include Hauksson (1981), Thomas (1988), and Toutain and Baubron (1999). One feature of these compilations is that the reported anomalies show the same type of magnitude-distance distributions as the hydrological responses we have seen so far – if the signals are real, there appears to be a threshold distance that scales with magnitude.

9.4.1 China: Haicheng, 1975 and Tangshan, 1976

The most celebrated and first (indeed only) prediction of a large earthquake was the 1975 M 7.3 Haicheng earthquake in China. Based in part on hundreds of hydrological anomalies an imminent prediction was made. Evacuations and preparations in Haicheng, with a population of about 1 million, contributed in part to the modest number of casualties, just over 2000. The prediction correctly identified the location, though not the precise time, of the event, and the magnitude was underestimated (Wang et al., 2006).

One and a half years later, the 1976 M 7.8 Tangshan earthquake occurred. Figure 9.2 shows the distribution of anomalies and time histories of radon concentration, groundwater level, land level and electrical resistivity in the region around Tangshan before and after the earthquake (from Ma et al., 1990). The fact that no prediction was issued, despite the abundance of potentially precursory anomalies,

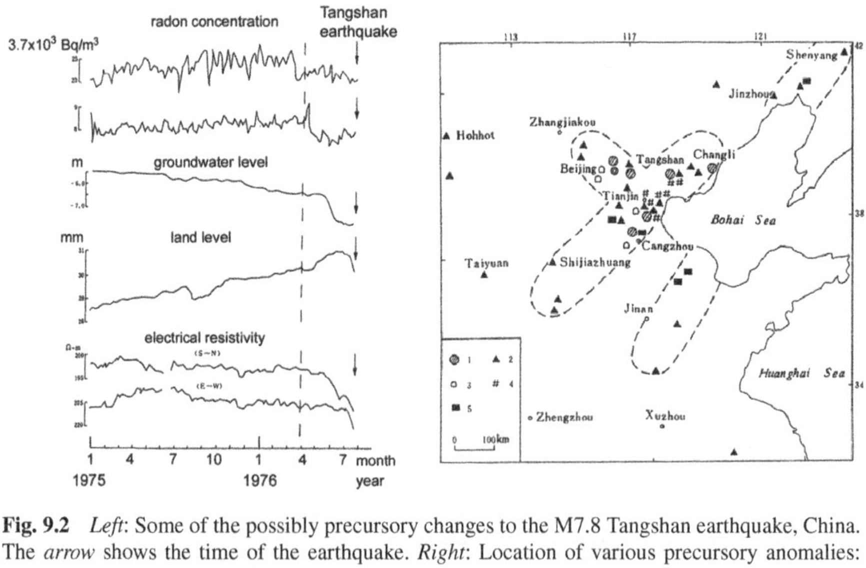

Fig. 9.2 *Left*: Some of the possibly precursory changes to the M7.8 Tangshan earthquake, China. The *arrow* shows the time of the earthquake. *Right*: Location of various precursory anomalies: resistivity (1), radon (2), land level (3), groundwater level (4), anomalies in oil wells (5) (From Ma et al., 1990)

highlights the difficulty in making predictions in a heavily populated area based on the existing data. Casualties from this earthquake exceeded 240,000.

9.4.2 *Kobe, Japan, 1995*

Following the 1995 M 7.2 Kobe earthquake several papers reported precursory changes in the concentrations of radon, chlorine, and sulfate ions in groundwater (e.g., Tsunogai and Wakita, 1995; Igarashi et al., 1995) and in groundwater level (King et al., 1995). The hydrogeochemical changes could be identified by analyzing bottled and dated spring water (Tsunogai and Wakita, 1995). Figure 9.3 shows a gradual increase in chloride concentration that begins 7 months before the earthquake. The initiation of these changes coincides with a 'drastic' increase in strain measured 5 km away from the well (Tsunogai and Wakita, 1995). This coincidence supports a broader tectonic origin of the pre-earthquake changes. However, whether the deformation responsible for hydrogeochemical changes and strain is connected to the later Kobe earthquake is difficult to evaluate. Given the length of the proposed precursory signal, a longer time series of measurements would be useful for establishing the uniqueness of the recorded changes.

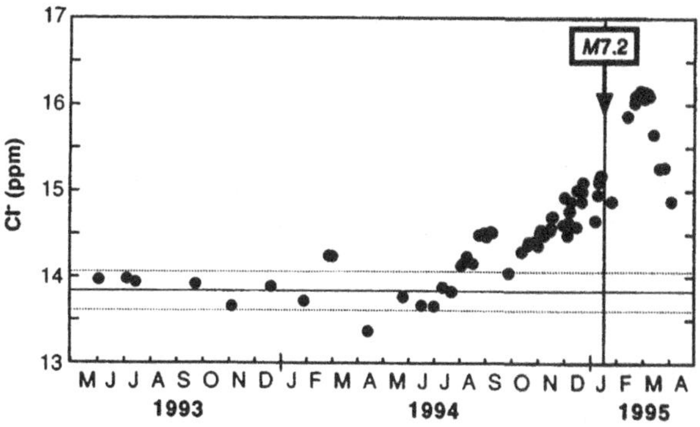

Fig. 9.3 Change in chloride concentration in bottled groundwater 20 km from the epicenter of the 1995 M7.2 Kobe earthquake , Japan. Time of the earthquake is shown by the *vertical line*. *Dashed horizontal lines* show one standard deviation of measurements prior to July 1994 (From Tsunogai and Wakita, 1995)

9.4.3 *Nankaido, Japan, 1946*

A few days prior to the 1946 M 8.3 Nankaido earthquake in Japan, water levels in some wells reportedly fell by more than 1 meter and some wells went dry (Sato, 1982). Linde and Sacks (2002) show that the pre-seismic deformation (observations reviewed in Roeloffs, 2006) can be explained by aseismic slip along the subduction interface. This area is now intensively monitored with 1200 continuous GPS stations. This data (Ozawa et al., 2002) along with leveling and tide-gauge data document other aseismic slip events (importantly, not followed by large earthquakes) in the region. Multiple and large aseismic events highlight the caution that a correlation of strain and hydrologic changes does not necessarily reflect deformation leading directly to a major earthquake, but possibly document events that remain purely aseismic.

The 1946 event was preceded by the 1944 M 8.2 Tonankai event, creating some ambiguity about whether the reported changes are responses or premonitory. Measured hydrological changes can lag behind tectonic strains (e.g., Ben-Zion et al., 1990) because of the time required for pore pressure diffusion.

9.4.4 Kettleman Hills, California, 1985

Three days before the 1985 M 6.1 Kettleman Hill earthquake, Roeloffs and Quilty (1997) found a gradual, anomalous rise in water level of about 3 cm in 2 of 4 wells in the nearby Parkfield area. These changes are shown in Fig. 9.1. Barometric pressure changes and rainfall cannot explain these changes. One of these two wells exhibited several similar changes that were not followed by earthquakes. In the second, however, the documented increase was unique during the 5 year monitoring period. Figure 9.1 shows that the sign of these possible precursory changes is opposite to the coseismic change implying that they are not caused by accelerating pre-seismic slip.

This observation was included in the IASPEI Preliminary List of Significant Precursors (Wyss and Booth, 1997). Nevertheless, important questions remain. What caused the anomalies? Why are they not recorded everywhere?

9.4.5 Chi-Chi, Taiwan, 1999

Some of the reported hydrologic ‘precursors’ to the 1999 Chi-Chi earthquake in Taiwan are interesting because they serve as good examples to illustrate some pitfalls in declaring hydrologic signals as precursors. The earthquake occurred at 1:47 a.m., September 21, local time. 59 of the 157 monitoring wells showed stepwise changes in groundwater level between the hour of 11 p.m., September 20, and 1 a.m., September 21. In other words, these records showed stepwise changes in groundwater level 1–3 h before the earthquake. If true and repeated for other earthquakes, these would be ideal precursors. Careful examination and verification of the clock of the recording instruments in the field and inspection of the information management process (Chia et al., 2000), however, necessitated a readjustment of the time-axis of the entire groundwater-level records. After corrections were made, all the ‘precursors’ turned out to be co-seismic responses.

A second example of possible misidentification of precursors is illustrated in Fig. 9.4. Stepwise changes on the groundwater-level records from four wells near the epicenter of the Chi-Chi earthquake were identified two weeks before the earthquake. These changes, however, turn out to be due to readjustment of the recording

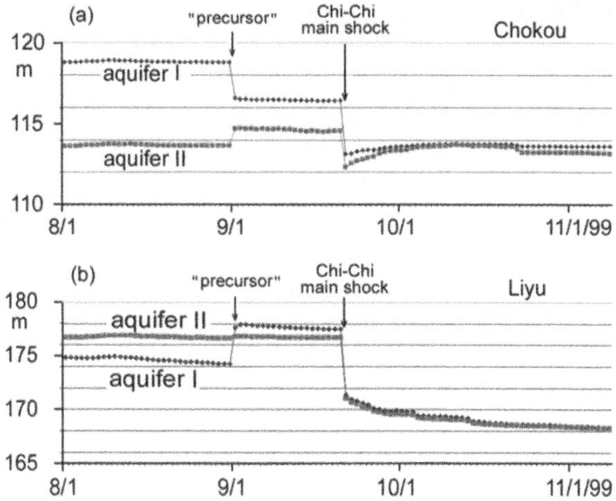

Fig. 9.4 Incorrectly identified groundwater level precursors to the 1999 M 7.3 Chi-Chi earthquake in Taiwan. The step a few weeks before the Chi-Chi earthquake is due to readjustment of the water level monitoring system

instruments (Y. Chia, 2007, personal communication). Both examples highlight the importance of (1) using a common time-base for the hydrologic and seismic records, and (2) marking all man-made instrumental changes as part of the hydrologic records.

Another reported possible precursor to the Chi-Chi earthquake is a change in the spectral characteristics of water level fluctuations in some wells in the month preceding the earthquake compared with those 2 and 3 months before the earthquake (Gau et al., 2007). This is not a compelling comparison as the amount and character of precipitation also changed. As discussed in Sect. 9.3 and Roeloffs (1988) the full range of relevant environmental factors must be considered. Given the long term memory and variability of hydrogeological systems, time series analysis should be undertaken for more than three months to assess the reliability of the analysis techniques in isolating seasonal effects, long term trends, and irregular variations (where any precursory signals would appear). A longer analysis could also identify the uniqueness of the reported precursory change – an essential attribute of any precursor.

Song et al. (2006) analyzed the composition of water at hot and artesian springs in Taiwan. Large, reversible anomalies in Cl^- or SO_4^{2-} were identified over a few year period. At the hot springs, anomalies precede earthquakes; however, anomalies do not exist before all earthquakes, and there is no correlation between the intensity of the shaking and the occurrence of precursory anomalies. Moreover, some anomalies are not followed by earthquakes. The artesian springs document postseismic changes, but these do not occur for all earthquakes and the occurrence of a response does not seem to be correlated with the intensity of shaking. Despite these severe limitations, Song et al. (2006) nevertheless claim that these springs are possibly ideal sites for recording precursors.

9.4.6 Kamchatka, 1992

Long term hydrogeochemical records are available in Kamchatka, an area with many large earthquakes. Biagi et al. (2006) illustrate a clear postseismic response in a spring following a M 7.1 earthquake about 100 km away (Fig. 9.5) and show that following this earthquake the spectral characteristics of the hydrogeochemical variations change, with an increase in short period variability. Biagi et al. (2006), expanding on Biagi et al. (2000), further claim that variations in other components in particular H_2 and CO_2 are precursory – their amplitude fluctuations decrease after the earthquake. We offer an alternative explanation for these changes and instead propose that they too are postseismic changes – the earthquake-created changes in hydraulic connectivity that leads to the changes shown in Fig. 9.5 are also responsible for the characteristic changes in H_2 and CO_2.

With a long record of hydrogeochemical monitoring and many earthquakes, Kamchatka offers an opportunity to test approaches to identifying precursors. Kingsley et al. (2001) identified as precursory, any signals that exceed 3 standard deviations of the mean and that are seen at the same time (within 7 days) in at least 2 measurements. With this criterion, they identify 8 precursors (anomalies within 158 days of the earthquake) and 3 failures (anomalies not followed by earthquakes) for a time period with 5 large (magnitudes between 6.9 and 7.3) earthquakes. With

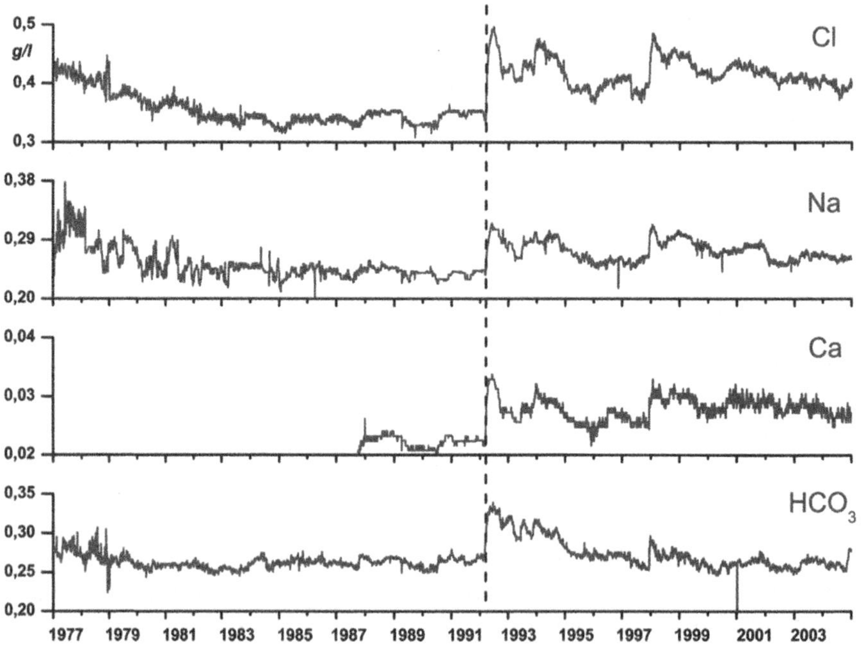

Fig. 9.5 Change in water composition at a spring in Katchatka. Water is sampled every 3 days. The time of a M 7.1 earthquake about 100 km distant is shown by the *vertical dashed line*. There is a clear postseismic response (Figure from Biagi et al., 2006)

a more restrictive criterion that anomalies are confined to ion data alone, Biagi et al. (2001) identify 3 anomalies, all of which are followed by earthquakes (the three closest large earthquakes to the wells). Examining their data (Figure 2 in Biagi et al., 2001), however, shows that there are correlated anomalies slightly smaller than 3 standard deviations that are not followed by earthquakes. Moreover, as the 5 large earthquakes occurred within less than a 5 year period, and correlated anomalies (greater than 3 sigma) occur every year or so, we thus expect that roughly half of identified precursory anomalies would fall within the 158 day time window simply by chance. Once again, we are left with several questions: What caused these anomalies? Why are some wells (apparently) more sensitive? What is the statistical significance of the anomalies?

9.4.7 Pyrenees, France, 1996

Toutain et al. (1997) analyzed the composition of bottled and dated spring water, as done following the 1995 Kobe earthquake (Tsunogai and Wakita, 1996), to document the pre- and postseismic response of groundwater to a M 5.2 earthquake in the French Pyrenees. The spring is located 29 km from the epicenter. As shown in Fig. 9.4, about 5 days before the earthquake the chloride concentration increased by about 40%, an increase much larger than the standard deviation of pre-seismic values (at least over the 200 days analysed). The high chloride values persisted for about another 5 days and then returned to 'normal'. Poitrasson et al. (1999) documented a lead anomaly in the same waters, also shown in Fig. 9.6. The lead anomaly has a shorter duration and is more than 10 times background values. The lead isotope changes suggest an anthropogenic source.

Toutain et al. (1997) propose that a small amount of chloride-rich water was injected into the aquifer feeding the springs – measured changes reflect mixing of previously isolated waters. The lead anomaly is not consistent with the possible sources for the chloride anomaly (Poitrasson et al., 1999), implying a third source of water. It is not clear why the start and end of the documented changes are so abrupt because dispersion should lead to more gradual changes, especially during the postseismic period.

9.4.8 Reservoir Induced Seismicity, Koyna, India

Chadha et al. (2003) report on an experiment to search for precursors to the reservoir-induced earthquakes near the Koyna and Warna reservoirs, India. The project involved drilling 19 wells for monitoring purposes. In addition to coseismic water level changes, Chadha et al. (2003) identify small, centimeter-scale, changes in water levels over periods of days to many days before earthquakes with magnitudes between 4.3 and 5.2 and within distances of 24 km.

Figure 9.7 shows an example of two of the premonitory changes, including the raw data and barometric pressure. The coseimic signal and possible precursory

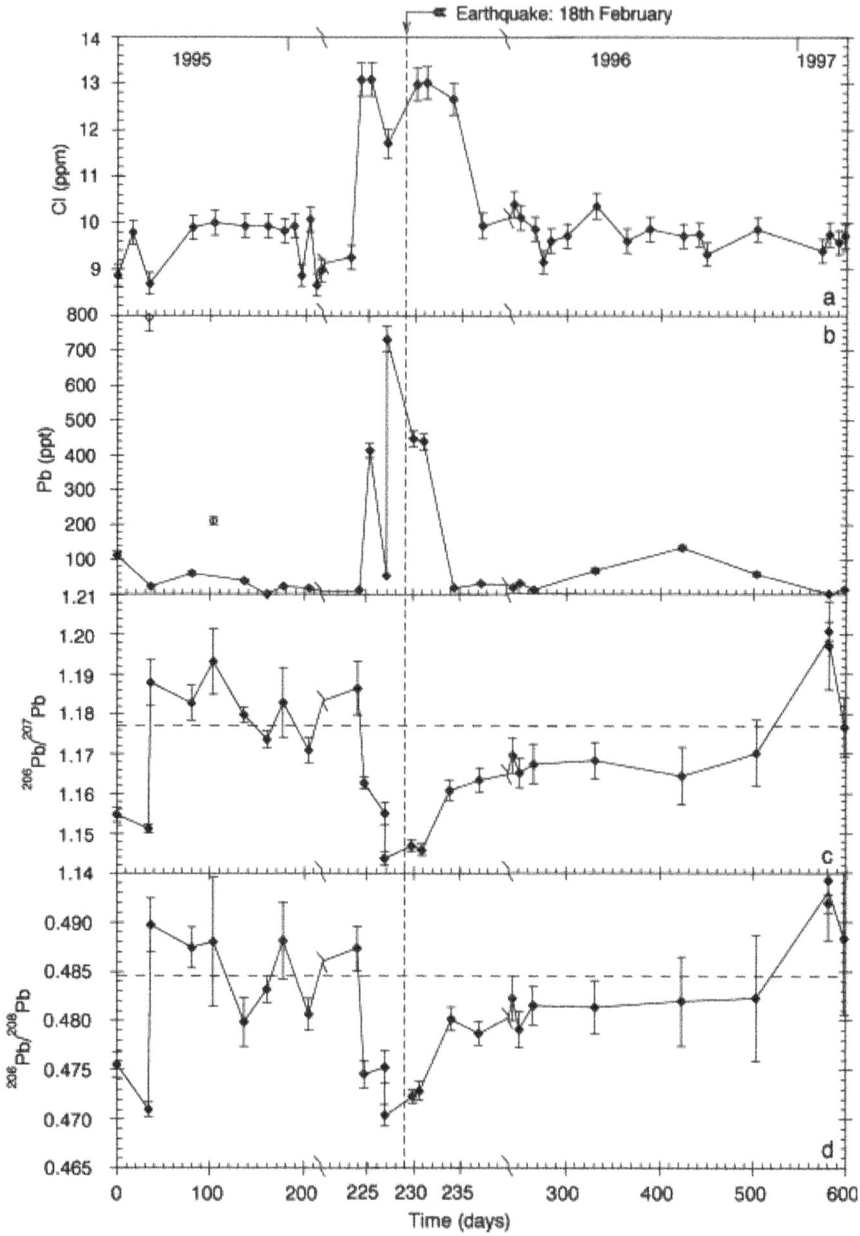

Fig. 9.6 Chloride and lead anomalies identified a posteriori from bottled waters. The time of a M 5.2 earthquake is shown by the *vertical dashed line*. The spring is located 29 km from the epicenter (From Poitrasson et al., 1999)

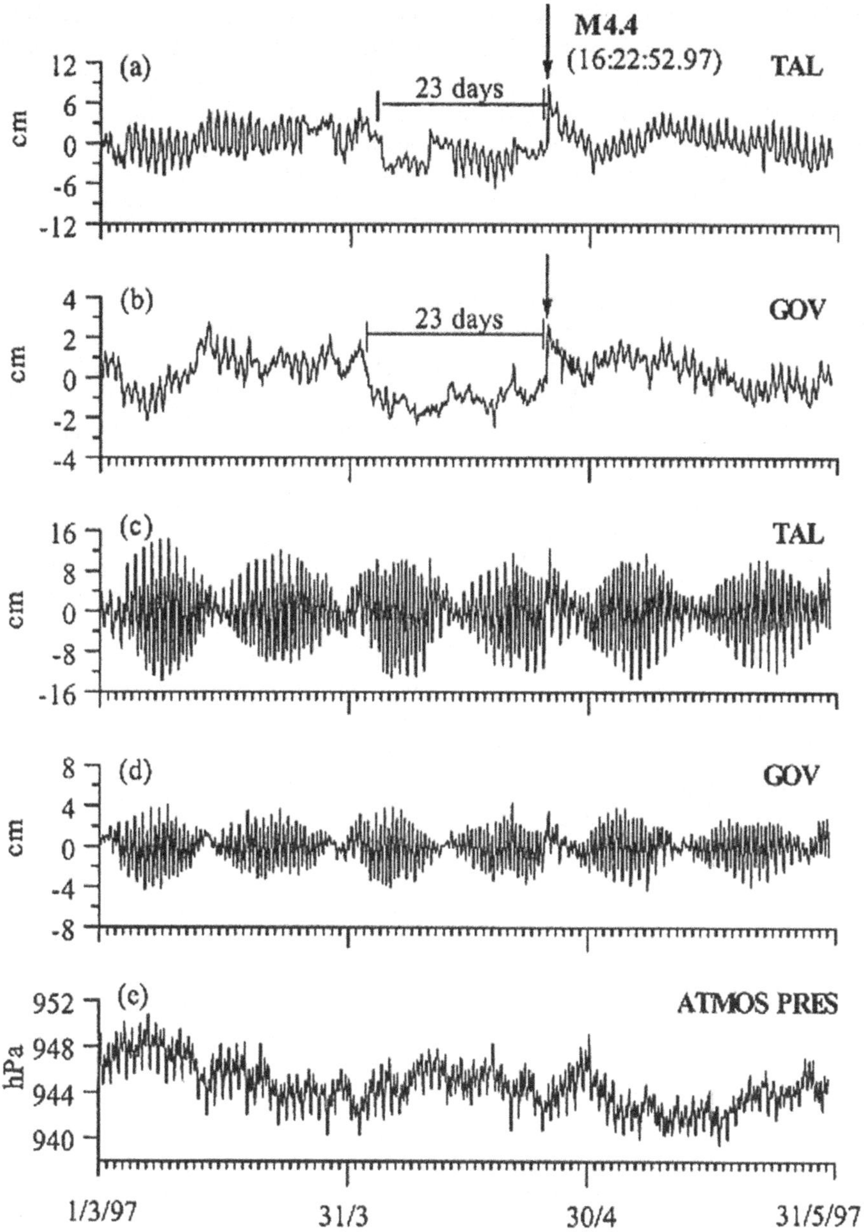

Fig. 9.7 Water level in two wells (TAL and GOV) over a 3 months period. **a** and **b** show water level after removing tides and barometric pressure effects (shown in **e**). **c** and **d** show raw water level records. The time of a M 4.4 earthquake that occurred 3 km from the wells is shown with the *arrow*. Chadha et al. (2003) claim that the 23 day period before this event is a precursory anomaly (From Chadha et al., 2003)

anomalies are dwarfed by the response to tides and barometric pressure. This study does not address the uniqueness of the proposed precursor anomalies. Inspection of Fig. 9.5 and other figures in this paper shows that similar anomalies occur and are not followed by earthquakes. Other questions remain about these purported precursors: Why don't all wells record the same anomalies? Why is the time duration of the anomalies different from earthquake-to-earthquake?

9.4.9 Calistoga Geyser, California

As discussed in Chap. 7, geysers are especially sensitive to small earthquake-generated strains. Silver and Vallette-Silver (1992) analyzed 18 years of eruptions at the Old Faithful, Calistoga, California geyser. During this period they documented three clear responses to regional earthquakes, as manifested in changes in the interval between eruptions (IBE, the most common measure, as discussed in Chap. 7, of geyser response) or the distribution of IBE. Two earthquakes caused an increase in IBE. The third caused a change in the mode of eruption, from a single IBE to multiple IBEs. These three earthquakes are consistent with a magnitude-distance threshold similar to other hydrological responses (Fig. 10.1).

Silver and Valette-Silver (1992) also propose that there are precursory changes in IBE that begin days before these three regional earthquakes. The data in this paper, however, clearly shows many features similar to the proposed precursory changes that were not followed by earthquakes. We believe that the statistical analysis in this paper significantly underestimates the number of times the IBE changes character, by perhaps 1–2 orders of magnitude, over the monitored period.

9.4.10 Precursory Changes in Spring Temperature

In Chap. 6 we discussed the co- and postseismic changes in temperature of a spring on the Izu Penisula, Japan (Mogi et al., 1989). At this spring, step-like increases in temperature of 1–2°C typically accompany regional earthquakes. Following earthquakes, temperature decreases approximately linearly. Also of note is a correlation of temperature changes with tides and barometric pressure, with magnitudes up to 0.5°C. Mogi et al. (1989) attributed these trends to be the result of unblocking followed by gradual resealing of fractures.

For a small number of earthquakes, as many as 5 (see Fig. 6.2), there are abnormal changes in temperature, defined as changes that are not coseismic, that do not follow the linear trend of decreasing temperature, and do not appear to be related to tides and weather. Mogi et al. (1989) referred to such abnormal changes as precursory. The changes occur between 3 days and 10 months before regional earthquakes. In one case, the precursory changes are coincident with a regional earthquake swarm.

As with other claimed precursors, there is no obvious predictive feature – the abnormal signals differ in form, timing, and do not always occur. Figures 7 and 9 in Mogi et al. (1989) also show abnormal changes not followed by earthquakes.

One possible explanation for the abnormal changes is that they are in fact responses to tectonic events – the 'precursory' response coincident with a regional earthquake swarm being an example. It is now recognized that there is a wide range of earthquake phenomena, particularly in subduction zone settings, in which slip does not only generate regular earthquakes. These events differ in the duration of the slip event, which can range from seconds for very-low frequency earthquakes (e.g., Ito et al., 2007), to hours for slow earthquakes (e.g., Linde et al., 1996) to days for slow-slip events (e.g., Hirose and Obara, 2005) to many months for silent earthquakes (e.g., Dragert et al., 2001; Ozawa et al., 2002; Kostoglodov et al., 2003). Such events are common in Japan and other subduction zones (Ide et al., 2007), but also occur along strike-slip faults such as the San Andreas in California (e.g., Linde et al., 1996) and at volcanoes (e.g., Segall et al., 2006).

If the 'precursory' changes reported by Mogi et al. (1989) are in fact responses to slower slip events than regular earthquakes, it suggests that the changes are more sensitive to the magnitude of strain rather than dynamic strains. At the same time, this also implies that 'precursors' are not useful for forecasting as not all slow earthquakes are followed by regular (and damaging) earthquakes.

9.5 Outlook

There are many retrospective reports of hydrological changes preceding earthquakes that appear to have no other obvious explanation. In very few cases, however, are the criteria met that are needed for critical evaluation – those listed in Sect. 9.3 and Roeloffs (1988). To identify these changes as precursory in a useful way also requires a criterion for distinguishing them from non-precursors before the actual earthquake occurs. Given the lack of success in using hydrological and other anomalies to predict earthquakes (including all three desired features: size, location and date) it is not surprising that earthquake prediction is not the focus of modern seismology. Readers may be surprised by our skepticism about most reported precursors and our critical assessment of the observations and data analysis. However, extraordinary claims require extraordinary proof (if not at least attention to detail); the ability to predict earthquakes certainly qualifies as an 'extraordinary claim'.

Hydrological precursors to earthquakes, if they exist, can be thought of as being a subset of a broad range of transient phenomena that includes silent and slow earthquakes, transient creep, episodic tremor and slip, and swarms (Bernard, 2001). Such transient phenomena occur more often and provide more measurement opportunities. Consequently, their study may prove more insightful about earthquake initiation and the types and origins of possible hydrological precursors or hydrologic phenomena that can be mistaken as precursors.

Multiparametric monitoring is particularly important both for identifying spurious anomalies and understanding the origin of hydrological changes. Combined deformation and water level measurements have proven useful to understand the spatio-temporal relationship between transient hydrological changes and deformation (e.g., Ben-Zion et al., 1990) and to support the identification of hydrological precursors (Roeloffs and Quilty, 1997).

Although we may still be far from achieving a complete understanding of the underlying mechanisms of the various earthquake-related anomalies that are reported in the literature, there remain significant monitoring efforts. A negative result, such as the absence of precursory signals at the multiparametric and densely monitored Parkfield, California site (Bakun et al., 2005), may frustrate the effort to predict earthquakes, but provides important and useful constraints on models of rupture initiation and other tectonic processes that lead up to earthquakes.

References

Aggarwal, Y.P., L.R. Sykes, J. Armbruster, and M.L. Sbar, 1973, Premonitory changes in seismic velocities and prediction of earthquakes, *Nature, 241*, 101.

Bakun, W.H., B. Aagard, B. Dost, W.L. Ellsworth, J.L. Hardebeck, R.A. Harris, C. Ji, M.J.S. Johnston, J. Langbein, J.J. Lienkaemper, et al., 2005, Implications for prediction and hazard assessment from the 2005 Parkfield earthquake, *Nature, 437*, 969–974.

Ben-Zion, Y., T.L. Henyey, P.C. Leary, and S.P. Lund, 1990, Observations and implications of water level and creepmeter anomalies in the Majave segment of the San Andreas fault, *Bull. Seism. Soc. Am., 80*, 1661–1676.

Bernard, P., 2001, From the search of 'precursors' to the research on 'crustal transients', *Tectonophysics, 338*, 225–232.

Biagi, P.F., L. Castellana, A. Minafra, G. Maggipinto, A. Ermini, O. Molchanov, Y.M. Khatkevich, and E.I. Gordeev, 2006, Groundwater chemical anomalies connected with the Kamchatka earthquake ($M = 7.1$) on March 1992, *Nat. Hazards Earth Syst. Sci., 6*, 853–859.

Biagi, P.F., A. Ermini, S.P. Kingsley., Y.M. Khatkevich, and E..I. Gordeev, 2000, Groundwater ion content precursors of strong earthquakes in Kamchatka (Russia), *PAGEOPH, 157*, 302–320.

Biagi, P.F., R. Piccolo, A. Ermini, Y. Fujinawa, S.P. Kingsley, Y.M. Khatkevich, and E.I. Gordeev (2001) Hydrogeochemical precursors of strong earthquakes in Kamchatka: further analysis, *Nat. Hazards Earth Syst. Sci., 1*, 9–14.

Bolt, B.A., and C.-Y. Wang, 1975, The present status of earthquake prediction, *Crit. Rev. Solid State Mater. Sci., 5*, 125–151.

Brace, W.F., J. Paulding, and C. Scholz, 1966, Dilatancy in the fracture of crystalline rocks, *J. Geophys. Res., 71*, 3939–3953.

Chadha, R.K., A.P. Pandey, and H.J. Kuempel, 2003, Search for earthquake precursors in well water levels in a localized seismically active area of reservoir triggered earthquakes in India, *Geophys. Res. Lett., 30*, 1416, doi:10.1029/2002GL016694.

Chia, Y., Y.S. Wang, H.P. Wu, C.J. Huang, C.W. Liu, M.L. Lin, and F.S. Jeng, 2000, Changes of ground water level in response to the Chi-Chi earthquake. In: *International Workshop on Annual Commemoration of Chi-Chi earthquake*, 317–328, National Center for Research on Earthquake Engineering, Taipei.

Claesson, L., A. Skelton, C. Graham, C. Dietl, M. Morth, P. Torssander, and I. Kockum, 2004, Hydrogeochemical changes before and after a major earthquake. *Geology, 32*, 641–644.

Claesson, L., A. Skelton, C. Graham, and M. Morth, 2007, The timescale and mechanisms of fault sealing and water-rock interaction after an earthquake, *Geofluids, 7*, 427–440.

Dragert, H., K. Wang, and T.S. James, 2001, A silent slip event on the deeper Cascadia subduction interface, *Science, 292*, 1525–1528.

Gau., H.S., T.-C. Chin, J.-S. Chen, and C.-W. Liu, 2007, Time series decomposition of groundwater level changes in wells due to the Chi-Chi earthquake in Taiwan: A possible hydrological precursor to earthquakes, *Hydrol. Proc., 21*, 510–524.

Hartmann, J., and J.K. Levy, 2005, Hydrogeological and gasgeochemical earthquake precursors – A review for application, *Nat. Hazards, 34*, 279–304.

Hauksson, E., 1981, Radon content of groundwater as an earthquake precursor: Evaluation of worldwide data and physical basis, *J. Geophys. Res., 86*, 9397–9410.

Hirose, H., and K. Obara, 2005, Repeating short- and long-term slow slip events with deep tremor activity, around the Bungo channel region, southwest Japan, *Earth Planets Space, 57*, 961–972.

Ide, S., G.C. Beroza, D.R. Shelly, and T. Uchide, 2007, A scaling law for slow earthquakes, *Nature, 447*, 76–79.

Igarashi, G., and H. Wakita, 1995, Geochemical and hydrological observations for earthquake prediction in Japan, *J. Phys. Earth, 43*, 585–596.

Igarashi, G., S. Saeki, N. Takahata, K. Sumikawa, S. Tasaka, Y. Sasaki, M. Takahasi, and Y. Sano, 1995, Groundwater radon anomaly before the Kobe earthquake in Japan, *Science, 269*, 60–61.

Ito, Y., K. Obara, K. Shiomi, S. Sekine, and H. Hirose, 2007, Slow earthquakes coincident with episodic tremors and slow slip events. *Science, 315*, 503–506.

King, C.-Y., 1980, Episodic radon changes in subsurface soil gas along active faults and possible relation to earthquakes, *J. Geophys. Res., 85*, 3065–3078.

King, C.-Y., D. Basler, T.S. Presser, W.C. Evans, L.D. White, and A. Minissale, 1994, In search of earthquake-related hydrologic and chemical changes along the Hayward fault, *Appl. Geochem., 9*, 83–91.

King, C.-Y., N. Koizumi, and Y. Kitigawa, 1995, Hydrogeochemical anomalies and the 1995 Kobe earthquake, *Science, 269*, 38–39.

Kingsley, S.P., P.F. Biagi, A. Ermini, Y.M. Khatkevich, and E.I. Gordeev, 2001, Hydrogeochemical precursors of strong earthquakes: A realistic possibility in Kamchatka, *Phys. Chem. Earth, 26*, 769–774.

Kondratenko, A.M. and Nersesov, I.L., 1962, Some results of the study of change in the velocity of longitudinal waves and relation between the velocities of longitudinal and transverse waves in a focal zone, *Trudy Insr. Fiziol. Zemli Acad. Nauk. USSR,* 25130.

Kostoglodov, V. et al., 2003, A large silent earthquake in the Guerrero seismic gap, Mexico, *Geophys. Res. Lett., 30*, doi:10.1029/2003GL017219.

Linde, A.T., and I.S. Sacks, 2002, Slow earthquakes and great earthquakes along the Nankai trough, *Earth. Planet. Sci. Lett., 203*, 265–275.

Linde, A.T., M.T. Gladwin, M.J.S. Johnston, R.L. Gwyther, and R.G. Bilham, 1996, A slow earthquake sequence on the San Andreas fault. *Nature, 383*, 65–67.

Lockner, D.A., and N.M. Beeler, 2002, Rock failure and earthquakes. In: W.H.K. Lee, et al. (eds.), *International Handbook of Earthquake and Engineering Seismology*, pp. 505–537, San Diego: Academic Press.

Ma, Z., Z. Fu, Y. Zhang, C. Wang, G. Zhang, and D. Liu, 1990, *Earthquake Prediction: Nine Major Earthquakes in China (1966–1976)*, pp. 332, Beijing: Seismological Press.

Maeda, K., and A. Yoshida, 1990, A probabilistic estimation of the appearance times of multiple precursory phenomena, *J. Phys. Earth, 38*, 431–444.

McEvilly, T.V., and L.R. Johnson, 1974 Stability of P and S velocities from central California quarry blasts, *Bull. Seism. Soc. Am., 64*, 343.

Mogi, K., H. Mochizuki, and Y. Kurokawa, 1989, Temperature changes in an artesian spring at Usami in the Izu Peninsula (Japan) and their relation to earthquakes, *Tectonophysics, 159*, 95–108.

Niu, F., P. Silver, T. Daley, X. Cheng, and E. Majer (2008) Preseismic velocity changes observed from active source monitoring at the Parkfield SAFOD drill site, *Nature, 454*, 204–208.

O'Connell R.J., and B. Budiansky, 1974, Seismic velocities in dry and saturated cracked solids, *J. Geophys. Res., 79*, 5412–5426.
Ozawa, S., M. Murakami, M. Kaidzu, T. Tada, T. Sagiya, Y. Hatanaka, H. Yarai, and T. Nishimura, 2002, Detection and monitoring of ongoing aseismic slip in the Tokai region, central Japan, *Science, 298*, 1009–1012.
Poitrasson, F., S.H. Dundaas, J.-P. Toutain, M. Munoz, and A. Rigo, 1999, Earth-related elemental and isotopic lead anomaly in a springwater, *Earth Planet. Sci. Lett., 169*, 269–276.
Quilty, E., and E. Roeloffs (1997) Water level changes in response to the December 20, 1994 M4.7 earthquake near Parkfield, California, *Bull. Seism. Soc. Am., 87*, 310–317.
Robinson, R., R.L. Wesson, and W.L. Ellsworth, 1974, Variation of P-wave velocity before the Bear Valley, California earthquake of 24 February, 1972, *Science, 184*, 1281.
Roeloffs, E.A., 1988, Hydrologic precursors to Earthquakes: A review, *PAGEOPH, 126*, 177–209.
Roeloffs, E.A., 2006, Evidence for aseismic deformation-rate changes prior to earthquakes, *Ann. Rev. Earth Planet. Sci., 34*, 591–627.
Roeloffs, E., and E. Quilty, 1997, Water level and strain changes preceding and following the August 4, 1985 Kettleman Hills, California, earthquake, *PAGEOPH, 149*, 21–60.
Sato, H., 1982, On the changes in the sea level at Tosashimuzu before the Nankaido earthquake of 1946, *J. Seismo. Soc. Japan, 35*, 623–626.
Scholz, C., L.R. Sykes, and Y.P. Aggarwal, 1973, Earthquake prediction: A physical basis, *Science, 181*, 803–810.
Segall, P., E.K. Desmarais, D. Shelly, A. Mikilius, and P. Cervelli, 2006, Earthquakes triggered by silent slip events on Kilauea volcano, Hawaii. *Nature, 442*, 71–74.
Semenov, A.N., 1969, Variations in the travel-time of transverse and longitudinal waves before violent earthquakes, *Izv. Acad. Sci. USSR Phys. Solid Earth.* (English Trans. No. 4), 245.
Silver, P.G., and N.J. Vallette-Silver, 1992, Detection of hydrothermal precursors to large northern California earthquakes, *Science, 257*, 1363–1368.
Song, S.R., W.Y. Wu, Y.L. Chen, C.M. Liu, H.F. Chen, P.S. Chan, Y.G. Chen, T.F. Yang, C.H. Chen, T.K. Liu, and M. Lee, 2006, Hydrogeochemical anomalies in the springs of the Chiayi Area in west-central Taiwan as possible precursors to earthquakes, *PAGEOPH, 163*, 675–691.
Sugisaki, R., 1978, Changing He/Ar and N_2/Ar ratios of fault air may be earthquake precursors, *Nature, 275*, 209–211.
Sugisaki, R., 1981, Deep-seated gas emission induced by the Earth tide: A basic observation for geochemical earthquake prediction, *Science, 212*, 1264–1266.
Sugisaki, R., and T. Sugiura, 1985, Geochemical indicator of tectonic stress resulting in an earthquake, central Japan, 1984, *Science, 229*, 1261–1262.
Thomas, D., 1988, Geochemical precursors to seismic activity, *PAGEOPH, 126*, 241–266.
Toll, N.J., and T.C. Rasmussen, 2007, Removal of barometric pressure effects and earth tides from observed water levels, *Ground Water, 45*, 101–105.
Toutain, J.P., M. Munoz, F. Poitrasson, and A.C. Lienard, 1997, Springwater chloride ion anomaly prior to a $M_L = 5.2$ Pyrenean earthquake, *Earth Planet. Sci. Lett., 149*, 113–119.
Toutain, J.P., and J. Baubron, 1999, Gas geochemistry and seismotectonics: A review, *Tectonophysics, 304*, 1–27.
Trique, N., P. Rochon, F. Perrier, J.P. Avouac, and J.C. Sabroux, 1999, Radon emanation and electric potential variations associated with transient deformation near reservoir lakes, *Nature, 399*, 137–141.
Tsunogai, U., and H. Wakita, 1995, Precursory chemical changes in groundwater: Kobe earthquake, Japan, *Science, 269*, 61–63.
Virk, H.S., and B. Singh, 1993, Radon anomalies in soil-gas and groundwater as earthquake precursor phenomena, *Tectonophysics, 227*, 215–224.
Wakita, H., 1996, Geochemical challenge to earthquake prediction, *Proc. Nat. Acad. Sci., 93*, 3781–3786.
Wakita, H., Y. Nakamura, and Y. Sano, 1988, Short-term and intermediate-term geochemical precursors, *PAGEOPH, 126*, 78–89.

Wang, H.F., 2000, *Theory of Linear Poroelasticity*, pp. 287, Princeton, NJ: Princeton University Press.

Wang, K, Q.-F. Chen, S. Sun, and A. Wong, 2006, Predicting the 1975 Haicheng earthquake, *Bull. Seism. Soc. Am., 96*, 757–795.

Whitcomb, J.H., J.D. Garmany, and D.L. Anderson, 1973, Earthquake prediction: Variation of seismic velocities before the San Fernando earthquake, *Science, 180*, 632–635.

Wyss, M., and D.C. Booth, 1997, The IASPEI procedure for the evaluation of earthquake precursors, *Geophys. J. Int., 131*, 423–424.

Chapter 10
Epilogue

Contents

In the previous chapters we discussed separately a wide variety of hydrologic responses to earthquakes, from liquefaction to triggered earthquakes. It is, however, not clear how these responses may relate to each other. In this final chapter, we first suggest a coherent framework that provides a integrated interpretation of the various hydrologic responses. We then suggest some directions for future research.

10.1 A General Framework

As shown in the previous chapters, to a first degree of approximation, the various hydrologic responses are scaled by earthquake magnitude M and hypocenter distance r. Thus they may be plotted together, as done in Fig. 10.1, to show the empirical relationships between r and M. While M and r are useful parameters for comparing and relating the various hydrologic responses, it would be more convenient if these two parameters are replaced with a single parameter to relate the responses. Ideally, this new parameter is a physical quantity that can be measured in laboratories, and hence could be used to tie field observations with laboratory measurements. The seismic energy density e, as defined in Chap. 2, may serve this purpose. It is the maximum energy available in the seismic wave train per unit volume to do work on rock or sediment at a given location, and may be easily estimated from the earthquake of magnitude M and the distance r to the earthquake source by the following empirical relation (Wang, 2007)

$$\log r = 0.48\,M - 0.33\log e(r) - 1.4 \tag{2.7}$$

C.-Y. Wang, M. Manga, *Earthquakes and Water*, Lecture Notes in Earth Sciences 114, DOI 10.1007/978-3-642-00810-8_10, © Springer-Verlag Berlin Heidelberg 2010

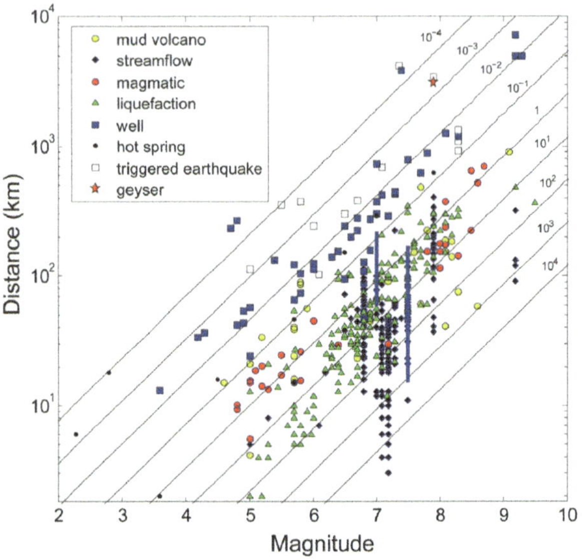

Fig. 10.1 Distribution of earthquake-induced hydrologic changes as functions of earthquake magnitude and epicentral distance. Also plotted are the contours of constant seismic energy density e, which is the seismic energy in a unit volume responding to the seismic wave train; it thus represents the maximum seismic energy available to do work at a given location during the earthquake. Note that some hot spring data shown here were not used by Mogi et al. (1989) in defining their linear M versus log r relationship for earthquake-induced changes in hot spring temperature (Fig. 6.2)

where r is in km and e in J/m^3. It may also be compared with laboratory results on the dissipated energy density required to initiate pore pressure change and liquefaction in saturated rock or sediment. The above relation shows that the contours of constant e plot as straight lines on a log r versus M diagram – these are added to Fig. 10.1. Since the seismic energy density is approximately proportional to the square of the peak ground velocity (Wang et al., 2006) which in turn is proportional to the dynamic strain (e.g., Brodsky et al., 2003), it provides a physically meaningful metric to relate and compare the various hydrologic responses.

Figure 10.1 shows that some hydrologic responses require much greater seismic energy density (e.g., liquefaction, mud volcanoes) than others (e.g., well level changes, earthquake triggering). On the other hand, each type of hydrologic response spans over four or more orders of magnitude of the seismic energy density values. Scatter in the hydrologic responses are expected for two reasons. First, if triggering is a threshold process, then for all distances up to the threshold we might expect triggering to be possible. Second, because the hydro-mechanical properties of rocks and sediments are highly variable, the range of sensitivity to seismic shaking may be large. Thus, for any specific hydrologic response to occur, less seismic energy density is required at sites underlain by sediments or rocks more sensitive to seismic disturbances than at sites underlain by less sensitive rocks or sediments. Without *a priori* knowledge of the seismic sensitivity of the rocks and sediments at a given site, we compare the different hydrologic responses by focusing on the threshold seismic energy density, i.e., the lower bound of the seismic energy density required to initiate a specific types of hydrological response in the most sensitive sediments or rocks. With this simplification in mind, Fig. 10.1 reveals that different hydrologic responses are bounded by different threshold seismic energy densities. Thus liquefaction, some mud volcanoes and streamflow increases are bounded by the contour with $e \sim 10^{-1}$ J/m^3, while groundwater level and hot-spring temperature may respond to $e \sim 10^{-4}$ J/m^3. It is important to note that most mud volcano eruptions are not triggered by earthquakes. The examples shown in Fig. 10.1 include only clearly identified triggered eruptions. Geysers have long been known to be particularly sensitive to earthquakes, as manifested by changes in the time interval between eruptions (Ingebritsen and Rojstaczer, 1993). Some geysers in the Yellowstone National Park, for example, have responded to seismic energy density as small as 10^{-3} J/m^3 (Fig. 10.1; see also Husen et al., 2004). Given the limited number of data, however, we are unable to confirm whether a M versus log r relationship may also apply to geysers.

Triggered seismicity also appears to be especially sensitive to seismic disturbances and may respond to e as small as 10^{-4} J/m^3 (Fig. 10.1; Brodsky and Prejean, 2005; Hill and Prejean, 2007). It is, however, important to note that the question whether all triggered seismicity is a hydrological phenomenon is a matter of active debate (e.g, Hill, 2008) and it is likely that some triggered earthquakes are not caused by earthquake-induced re-distribution of pore pressure. Regardless of a clear hydrologic connection, triggered earthquakes by large ($M > 9$) earthquakes may be global (e.g., West et al., 2005; Velasco et al., 2008) – consistent with the threshold limit for triggered earthquakes shown in Fig. 10.1.

Part of the differences in the threshold energy between different hydrological responses may be a result of incomplete data. For example, the data for hot spring temperature come from a single hot spring in Japan (Mogi et al., 1989) and there is only one data point that falls on the contour of $e = 10^{-4}$ J/m^3. Thus we cannot be certain if a M versus log r relationship may also apply at such low e. On the other hand, most other data summarized in Fig. 10.1 are abundant, come from a wide range of geological settings, and thus the differences in the threshold energy among the different hydrologic responses may be significant.

As discussed in Chap. 2, a great number of laboratory experiments by earthquake engineers have shown that a minimum dissipated energy density of 30 J/m^3 is required to cause liquefaction in sensitive soils (Green and Mitchell, 2004), which is more than 10^2 times greater than the seismic energy density at the threshold distance for liquefaction occurrence, i.e., 0.1 J/m^3 (Fig. 10.1; see also Wang, 2007). As also discussed in Chap. 5, under cyclic loading pore pressure begins to build up in various saturated sediments under a broad range of confining pressures when the shear strain exceeds a threshold of 10^{-4}. Combining this result with the damping ratios for sediments under cyclic loading, Wang and Chia (2008) showed that the dissipated energy density required to initiate undrained consolidation ranges from 0.1 to 10 J/m^3, which is more than 10^3 times greater than the seismic energy density at the threshold distance of sustained groundwater-level changes, i.e., 10^{-4} J/m^3 (Fig. 10.1). Thus undrained consolidation is expected to cause liquefaction only within the near field, and may initiate groundwater level changes only up to the intermediate field. Since the field occurrences of liquefaction and sustained pore-pressure changes far exceed these limits, different mechanisms are required to explain these more distant responses.

As also discussed earlier, several mechanisms besides undrained consolidation have been proposed to explain hydrologic responses, including static poroelastic strain associated with fault displacement (Wakita, 1975; Muir-Wood and King, 1993; Quilty and Roeloffs, 1997), dynamic strain associated with seismic waves (Manga and Wang, 2007) and enhanced permeability of the shallow crust (Mogi et al., 1989; Rojstaczer et al., 1995; Brodsky et al., 2003; Wang et al., 2004). At distances beyond the near field, static poroelastic strain is so small that it cannot easily account for the observed large amplitude hydrologic changes (Manga and Wang, 2007); furthermore, the model predicted the wrong sign of most observed groundwater-level changes during the 1999 Chi-Chi earthquake (Wang et al., 2001; Koizumi et al., 2004) and cannot explain the persistent groundwater level increases in the BV well in central California (Roeloffs, 1998) or the persistent streamflow increases in Sespe Creek (Manga et al., 2003), southern California, in response to multiple earthquakes of different mechanisms and orientations.

Dynamic strain by itself cannot lead to sustained hydrologic changes, but it can enhance permeability of the shallow crust and redistribute pore fluids by dislodging blockage from fractures. Mogi et al. (1989) first suggested that seismic waves may dislodge obstacles from hot spring passageways and thus to enhance permeability and to cause the coseismic increase in hot spring temperature (Chap. 6). The same mechanism was used to explain the sustained changes in groundwater level (Roeloffs, 1998; Brodsky et al., 2003; Wang and Chia, 2008) and the increase in stream discharge (Rojstaczer et al., 1995; Wang et al., 2004) after large earthquakes. Changes in the eruption frequency of geysers can also be caused by changes in permeability of the conduit and/or surrounding matrix (Ingebritsen and Rojstaczer, 1993). As the permeability of the conduit is very high, changes in the matrix permeability that governs conduit recharged are more likely (Ingebritsen and Rojstaczer, 1993; Manga and Brodsky, 2006). Elkhoury et al., (2006) showed, through analysis of the tidal response of groundwater level, that seismic waves can indeed significantly enhance the permeability of shallow crust and the magnitude of this enhancement increases with increased peak ground velocity, i.e., with increased seismic energy density. Other authors (Roeloffs, 1998) noticed that, at a given well, the amplitude of the sustained groundwater-level change increases in proportion to the increased peak ground velocity, i.e., seismic energy. This may be expected as greater seismic energy (more rigorous shaking) may clear up more blockage from fluid passageways, resulting in greater increase in permeability and more efficient redistribution of pore pressure. Taken together, it appears that enhanced permeability in the shallow crust during earthquakes may be a common mechanism for a broad spectrum of hydrologic responses that occur in the intermediate and far fields.

It remains to be shown that seismic waves with energy densities as small as $\sim 10^{-4}$ J/m^3 can still enhance permeability. Numerical simulation (Wiesner, 1999) show that, at low flow velocity, clay particles suspended in water form flocculated deposits which effectively fill fractures, blocking flow. Mechanically, such fluids are non-Newtonian and have a yield strength arising from the bonds among clay particles (Coussot, 1995). The yield strength is equivalent to a threshold energy density required to disrupt the bonds among clay particles, which is 10^{-3} J/m^3 at a few percent solid fraction for several different clays (Coussot, 1995). It may be even smaller under oscillatory seismic loading. Thus, for fracture porosity less than 10^{-1}, seismic waves with energy density as low as $\sim 10^{-4}$ J/m^3 may still break up clay networks in fractures to enhance groundwater flow. Increased flow may induce greater permeability (Wiesner, 1999), resulting in greater flow, etc. The 2002 M7.9 Denali earthquake, for example, enhanced groundwater flow in Iowa, some 5000 km away, to such an extent that colloidal particles were flushed from local aquifers to discolor well waters (http://www.igsb.uiowa.edu/gwbasics/Chapters/Wells%20Getting%20Water%20From%20the%20Ground.pdf).

Thus starting at a threshold energy density of $\sim 10^{-4}$ J/m^3, seismic waves may dislodge minute clogs from fractures to enhance permeability and to redistribute pore pressure, resulting in changes in groundwater level in the most sensitive wells, changes in hot spring temperature and triggered seismicity. Increasing seismic energy density may remove larger blockages from fractures to allow more efficient groundwater flow to cause noticeable groundwater-level changes in less sensitive wells and to change geysers' eruption frequency. Continued increase in pore pressure and removal of grains act to degrade the soil stiffness during each seismic cycle until the soil finally liquefies (Holzer and Youd, 2007). Since mud volcanoes triggered by earthquakes require the liquefaction or fluidization of the erupted material, it is natural to expect that earthquake-triggered mud volcanoes and liquefaction are created by the same process and thus constrained by similar threshold of seismic energy density (Fig. 10.1).

Earthquake-enhanced permeability must also occur in the near field, but its hydrologic effects in the near field may be obscured by those caused by undrained consolidation. In evaluating the site safety to earthquakes, earthquake engineers may need to consider the occurrence of pore-pressure change and liquefaction beyond the near field and to study other mechanisms for such changes, in addition to undrained consolidation.

10.2 Directions for Future Research

Much remains unexplored in the study of the hydrologic responses to earthquakes. In the following we highlight some directions for future research which may further our understanding of these responses.

(1) *Frequency effects*: As noted in Chap. 2, the results from field and laboratory studies on the dependence of liquefaction on frequency are apparently in conflict. On the one hand, laboratory results show little frequency-dependence of liquefaction (Yoshimi and Oh-Oka, 1975; Sumita and Manga, 2008); on the other hand, in situ evidence from seismically instrumented liquefaction sites show an association of liquefaction to low-frequency ground motions (e.g., Youd and Carter, 2005; Holzer and Youd, 2007; Wong and Wang, 2007). Future research, including *in situ*, laboratory, and theoretical work, is required to explain these differences.

(2) *Testing the enhanced permeability hypothesis*: As discussed earlier in this chapter, the threshold seismic energies for some hydrologic responses are extremely low (10^{-4} to 10^{-3} J/m^3). We hypothesized that these low threshold energies represent those required to enhance the crustal permeability by dislodging obstacles from pre-existing cracks and fractures (Wang and Chia, 2008). Elkhoury et al. (2006) showed that seismic waves can indeed increase the crustal permeability. But can this occur at seismic energies as low as 10^{-4} to 10^{-3} J/m^3? Future research, including *in situ*, laboratory, and theoretical work, is required to test this hypothesis.

(3) *Effects of rupture directivity and faulting style*: Since the distribution of seismic energy in the near field depends on the directivity of rupture and on the style of faulting, we may expect such dependency to show up in the distribution of the hydrologic responses. However, most published work so far has focused on earthquake magnitude and hypocenter distance as the only parameters to characterize the distribution of the hydrologic responses to earthquakes. Part of the reason is that these two parameters are the most easily available, especially for historical earthquakes. Given that dense networks of strong motion seismographs and hydrologic monitory wells are becoming widespread in earthquake countries, it is now relatively straightforward to study the dependency of hydrologic responses on the directivity of rupture and on the style of faulting. The result of such efforts can provide important constraints on the models of the mechanism for the hydrologic responses.

(4) *Site effects*: Another deficiency of using earthquake magnitude and hypocenter distance as the only parameters in characterizing the distribution of the hydrologic responses is a complete absence of the 'site effect'. Since site effects have a strong influence on the distribution of seismic energy, any attempt to make predictions regarding liquefaction or other hydrologic responses at a particular site must incorporate site-specific geologic data. Recent advances in numerical simulation of the regional propagation and attenuation of seismic waves from an earthquake source have allowed the inclusion of detailed geology of the modeled region, and thus the site effect at any point in this region is implicitly included in the simulated seismic energy density, at a resolution determined by the grid size used in the simulation. A comparison among the simulated seismic energy distribution, the distribution of soil types, and the energy criterion proposed in this book and related papers may provide a first-order estimate of

the liquefaction risk or other hydrologic hazards in the region during a potential earthquake on a particular fault.

(5) *More observations*: While it goes without saying that more observations and quantitative data are useful, there is a need for integrated hydrogeochemical, hydrological, temperature, and deformation measurements. As we discussed in Chap. 9 on earthquake precursors, limited sampling and short time series often limit the ability to test hypotheses and reliably identify hydrologic responses and precursors.

References

Brodsky, E.E., E. Roeloffs, D. Woodcock, I. Gall, and M. Manga, 2003, A mechanism for sustained water pressure changes induced by distant earthquakes, *J. Geophys. Res., 108*, doi:10.1029/2002JB002321.

Brodsky, E.E., and S.G. Prejean, 2005, New constraints on mechanisms of remotely triggered seismicity at Long Valley Caldera, *J. Geophys. Res., 110*, doi:10.1029/2004JB003211.

Coussot, P., 1995, Structural similarity and transition from Newtonian to non-Newtonian behavior for clay-water suspensions, *Phys. Rev. Lett., 74*, 3971–3974.

Elkhoury, J.E., E.E. Brodsky, and D.C. Agnew, 2006, Seismic waves increase permeability, *Nature, 411*, 1135–1138.

Green, R.A., and J.K. Mitchell, 2004, Energy-based evaluation and remediation of liquefiable soils. In: M. Yegian, and E. Kavazanjian (eds.), *Geotechnical Engineering for Transportation Projects, ASCE Geotechnical Special Publication, No. 126, Vol. 2,* 1961–1970.

Hill, D.P., and S.G. Prejean, 2007, Dynamic triggering. In: *Treatise on Geophysics*, G. Schubert editor, Vol. 4, pp. 293–320.

Hill, D.P. (2008) Dynamic stresses, coulomb failure, and remote triggering, *Bull. Seism. Soc. Am., 98*, 66–92.

Holzer, T.L., and T.L. Youd, 2007, Liquefaction, ground oscillation, and soil deformation at the Wildlife Array, California, *Bull .Seism. Soc. Am., 97*, 961–976.

Husen, S., R. Taylor, R.B. Smith, and H. Healser, 2004, Changes in geyser eruption behavior and remotely triggered seismicity in Yellowstone. National park produced by the 2002 M 7.9 Denali fault earthquake, Alaska, *Geology, 32*, 537–540.

Ingebritsen, S.E., and S.A. Rojstaczer, 1993, Controls on geyser periodicity, *Science, 262*, 889–892.

Koizumi N., W.-C. Lai, Y. Kitagawa, and Y. Matsumoto, 2004, Comment on "Coseismic hydrological changes associated with dislocation of the September 21, 1999 Chichi earthquake, Taiwan" by Min Lee et al. GRL, 31, L13603, doi:10.1029/2004GL019897.

Manga, M., E.E. Brodsky, and M. Boone, 2003, Response of streamflow to multiple earthquakes and implications for the origin of postseismic discharge changes, *Geophys. Res. Lett., 30*, doi:10.1029/2002GL016618.

Manga, M., and E. Brodsky, 2006, Seismic triggering of eruptions in the far field: Volcanoes and geysers, *Annu. Rev. Earth Planet. Sci., 34*, 263–291.

Manga, M., and C.-Y. Wang, 2007, Earthquake hydrology. In: *Treatise on Geophysics*, G. Schubert editor, Vol. 4, pp. 293–320.

Mogi, K., H. Mochizuki, and Y. Kurokawa, 1989, Temperature changes in an artesian spring at Usami in the Izu Peninsula (Japan) and their relation to earthquakes, *Tectonophysics, 159*, 95–108.

Muir-Wood, R., and G.C.P. King, 1993, Hydrological signatures of earthquake strain *J.Geophys.Res., 98*, 22035–22068.

Quilty, E.G., and E.A. Roeloffs, 1997, Water level changes in response to the December 20, 1994 M4.7 earthquake near Parkfield, California, *Bull. Seism. Soc. Am., 87*, 1018–1040.

Roeloffs, E.A., 1998, Persistent water level changes in a well near Parkfield, California, due to localand distant earthquakes, *J. Geophys. Res., 103*, 869–889.

Rojstaczer, S., S. Wolf, and R. Michel, 1995, Permeability enhancement in the shallow crust as a cause of earthquake-induced hydrological changes, *Nature, 373*, 237–239.

Sumita, I., and M. Manga, 2008, Suspension rheology under oscillatory shear and its geophysical implications, *Earth Planet. Sci. Lett., 269*, 467–476.

Velasco, A.A., S. Hernandez, T. Parsons, and K. Pankow, 2008, Global ubiquity of dynamic earthquake triggering. *Nature Geosci., 1*, 375–379.

Wakita, H., 1975, Water wells as possible indicators of tectonic strain, *Science, 189*, 553–555.

Wang, C.-Y., L.H. Cheng, C.V. Chin, and S.B. Yu, 2001, Coseismic hydrologic response of an alluvial fan to the 1999 Chi-Chi earthquake, Taiwan, *Geology, 29*, 831–834.

Wang, C.-Y., 2007, Liquefaction beyond the near field, *Seism. Res. Lett., 78*, 512–517.

Wang, C.-Y., C.-H. Wang, and M. Manga, 2004, Coseismic release of water from mountains – evidence from the 1999 ($M_W = 7.5$) Chi-Chi, Taiwan earthquake, *Geology, 32*, 769–772.

Wang, C.-Y., A. Wong, D.S. Dreger, and M. Manga, 2006, Liquefaction limit during earthquakes and underground explosions – implications on ground-motion attenuation, *Bull. Seis. Soc. Am., 96*, 355–363.

Wang, C.-Y., and Y. Chia, 2008, Mechanism of water-level changes during earthquakes: Near field versus intermediate field, *Geophys. Res. Lett., 35*, L12402, doi:10.1029/2008GL034227.

West, M., J.J. Sanchez, and S.R. McNutt, 2005, Periodically triggered seismicity at Mount Wrangell, Alaska, after the Sumatra earthquake, *Science, 308*, 1144–1146.

Wiesner, M.R., 1999, Morphology of particle deposits, *J. Environ. Engin., 125*, 1124–1132.

Wong, A., and C.-Y. Wang, 2007, Field relations between the spectral composition of ground motion and hydrological effects during the 1999 Chi-Chi (Taiwan) earthquake. *J. Geophys. Res., 112*, B10305, doi:10.1029/2006JB004516.

Yoshimi, Y., and H. Oh-Oka, 1975, Influence of degree of shear stress reversal on the liquefaction potential of saturated sand, *Soil. Found.* (Japan), *15*, 27–40.

Youd, T. L., and B. L. Carter, 2005, Influence of soil softening and liquefaction on spectral acceleration, *J. Geotech. Geoenviron. Eng., 131*, 811–825.

Appendices

Contents

C.-Y. Wang, M. Manga, *Earthquakes and Water*, Lecture Notes in Earth Sciences 114, DOI 10.1007/978-3-642-00810-8,

Appendix A: Notation

Symbol	Quantity
a	Constant in recession equation (2.1)
A	Recharge/unit area
A	Amplitude of an oscillation
b	Aquifer thickness
B	Skempton coefficient
c	Baseflow recession constant
c	Cohesive strength
c_s	Specific heat of solid phase
c_f	Specific heat of fluid phase
C	Concentration of dissolved constituent
D	Hydraulic diffusivity
D_w	Chemical diffusion coefficient in water
D_m	Chemical diffusion coefficient in porous material
D	Hydrodynamic dispersion
D_L	Longitudinal dispersion coefficient
D_T	Transverse dispersion coefficient
e	Energy density
E	Energy
f	Fluid content
h	Hydraulic head or groundwater level
H	Heat source/sink
k	Permeability
K	Hydraulic conductivity
K_h	Thermal conductivity
K	Drained bulk modulus
K_s	Solid grain bulk modulus
K_u	Undrained bulk modulus
L	Aquifer length
M	Earthquake magnitude
n	Porosity
n_e	Effective porosity
N	Number of cycles
P	Pressure
Pe_c	Compositional Peclet number
Pe_{th}	Thermal Peclet number
q	Discharge
Q_s	Source/sink of solute
r	Distance from earthquake epicenter
R	Maximum distance for a specified hydrologic response from the earthquake source
r_c	Inner radius of well casing
r_w	Radius of a well
S_s	Specific storage
S_y	Specific yield
t	Time
T	Temperature
T	Transmissivity
x	Position

(continued)

Symbol	Quantity
α	Biot-Willis coefficient
α_L	Longitudinal dispersivity
α_T	Transverse dispersivity
β	Compressibility
ε	Strain
γ	Shear strain
η	Phase of an oscillation
μ	Shear modulus
μ	Fluid viscosity
ρ_f	Fluid density
ρ_r	Solid (rock) density
σ	Stress
σ'	Effective stress
τ	Shear stress
τ	Characteristic time of an aquifer
σ_N	Normal stress
ν	Linear velocity
ω	Frequency of an oscillation

Notes:
(a) Bold symbols denote vector and tensor quantities.
(b) Subscripts on tensor quantities, e.g., σ_{ij}, denote components of the tensor.

Appendix B: Basic Equations for Groundwater Flow

B.1 Darcy's law

Henry Darcy's (1856) experiment and Darcy's law:

$$Q = \frac{KA(h_1 - h_2)}{L}, \tag{B.1}$$

$$q = -K\frac{dh}{dL} \tag{B.2}$$

where, referring to the following diagram, Q is the total discharge through a column of porous sediments or rock of cross-section area of A, q the discharge per unit area, i.e., Q/A, h the hydraulic head, L the distance between points 1 and 2, and K is the hydraulic conductivity of the sediments or rock.

Q has a dimension of L^3/T, and both q and K have a dimension of L/T, same as velocity. The transmissivity T of an aquifer of thickness b is defined as

$$T = bK. \tag{B.3}$$

Limitation of Darcy's law: valid only if the flow is laminar.

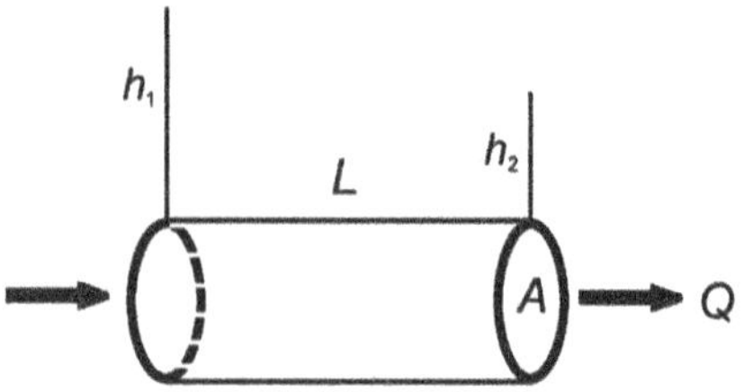

Fig. B.1 A-b.1 Schematics of Darcy's experiment to show sample dimensions (sample cross-section area A and length L), the hydraulic heads (h_1 and h_2) at the two ends of the sample, and the discharge Q through the sample

Reynold's number:

$$R_e = \frac{\rho_f \, q \, d}{\mu} \tag{B.4}$$

where ρ_f and μ are, respectively, the density and viscosity of the fluid and d is the average diameter of the pores. For porous flow, flow is laminar at $R_e < 5$. When $R_e \geq 5$, the flow deviates from laminar flow and become turbulent, and Darcy's law is no longer valid.

Relation between hydraulic conductivity and permeability

$$K = \frac{\rho_f \, g \, k}{\mu} \tag{B.5}$$

B.2 Porosity and Permeability

Definition of porosity: $n = V_v/V$, where V_v is the volume of pores and V is the bulk volume of sediments. n ranges from $\sim 10^{-2}$ for crystalline rocks, 10^{-1} for sedimentary rocks, and from 0.3 to 0.8 for unconsolidated sediments and soils.

Generally, porosity decreases with depth, largely as a result of pressure induced consolidation, and this burial-induced decrease is mostly irreversible. The rate of this depth, however, depends on the compressibility of the porous medium. For sedimentary basins, a commonly used empirical relation is Athy's 'law' (Athy, 1930),

$$n = n_o \, e^{-bz} \tag{B.6}$$

where n_o is the porosity at the surface, z is depth, and b an empirical constant.

For different earth media, permeability varies over 16 orders of magnitude. It generally depends on porosity. A commonly used equation relating permeability to porosity is the Kozeny-Carmen relation:

$$k = \frac{n^3}{5\,(1 - n)^2 \, s_o} \tag{B.7}$$

where s_o is the area of the solid surface exposed to the fluid per unit volume of matrix.

Given the various heterogeneities in the crust that become involved at different scales, permeability may be a function of the scale of the problem. Laboratory specimens have volumes $\ll 1\ m^3$. In situ measurements normally represent a volume from 10 m^3 to $10^5\ m^3$. For studies of regional flows, where direct measurement of permeability is often absent, permeability may sometimes be estimated from model 'fitting' to the regional hydrological data.

Permeability also generally decreases with depth, both due to a decrease of porosity and due to hydrothermal deposits. Several empirical relations have been proposed, including Manning and Ingebritsen's (1999) and Saar and Manga (2004).

In general, permeability is anisotropic and is represented by a tensor, i.e.,

$$\underline{K} = \begin{pmatrix} K_{xx} & K_{xy} & K_{xz} \\ K_{yx} & K_{yy} & K_{yz} \\ K_{zx} & K_{zy} & K_{zz} \end{pmatrix} \tag{B.8}$$

Thus Darcy's law takes the following general expression,

$$\begin{pmatrix} q_x \\ q_y \\ q_z \end{pmatrix} = \begin{pmatrix} K_{xx} & K_{xy} & K_{xz} \\ K_{yx} & K_{yy} & K_{yz} \\ K_{zx} & K_{zy} & K_{zz} \end{pmatrix} \begin{pmatrix} \partial h/\partial x \\ \partial h/\partial y \\ \partial h/\partial z \end{pmatrix} \tag{B.9}$$

Or, in vector form,

$$\boldsymbol{q} = -\underline{\boldsymbol{K}} \cdot \nabla h. \tag{B.10}$$

The average permeability of layered sediments or rocks is anisotropic even though the material within each layer is isotropic. In such case, the average permeability parallel to the bedding of the layered rocks is:

$$k = \sum_i k_i \left(\frac{b_i}{b_t}\right) \tag{B.11}$$

where k_i is the permeability of the *ith* layer. On the other hand, the average permeability normal to the bedding of layered rocks is:

$$k = \left(\frac{b_t}{\sum_i b_i / k_i}\right) \tag{B.12}$$

Permeability may gradually evolve through time due to ongoing geo-biologic processes, dissolution, precipitation and formation of clay minerals. The time scales for calcite dissolution is 10^4 to 10^5 yr and for silica precipitation is weeks to years (Ingebritsen et al., 2006). Permeability may also change suddenly during an earthquake, at time scales of one to tens of seconds.

B.3 Elements in a Groundwater System

A groundwater system may consist of several individual aquifers and aquitards. Aquifers are rock or sediment formations in which groundwater can move with relative ease. Aquitards are rock or sediment formations in which groundwater moves with difficulty. A confined aquifer is an aquifer that is confined on both the upper and lower boundaries by aquitards. An unconfined aquifer is an aquifers confined only on its lower boundary. Water table is the top of the saturated zone in the subsurface.

B.4 Driving Potential

The driving potential of groundwater flow is the hydraulic head, h, defined as the sum of the pressure head and the elevation head:

$$h = \frac{P}{\rho_f g} + z \tag{B.13}$$

B.5 The Continuum Approach

In the continuum approach, a representative elemental volume (REV) is the chosen in which the hydraulic properties are assumed uniform. This volume needs to be very much greater than the sizes of the individual pores and cracks so that the average properties are not dependent on the REV, but very much smaller than the domain of study.

B.6 The Groundwater Flow Equations

Combining the equation for the conservation of fluid mass

$$\frac{\partial (n \rho_f)}{\partial t} = -\nabla \cdot (\rho_f \boldsymbol{q}) \tag{B.14}$$

and Darcy's law $\boldsymbol{q} = -\underline{\boldsymbol{K}} \cdot \nabla h$, we have

$$\frac{\partial (n \rho_f)}{\partial t} = \nabla \cdot (\rho_f \underline{\boldsymbol{K}} \cdot \nabla h) \tag{B.15}$$

By orienting the coordinates along the principal directions of the $\underline{\boldsymbol{K}}$ tensor, all the off-diagonal elements in the conductivity matrix become zero and the diagonal elements take their principle values K_x, K_y and K_z. If K_x, K_y and K_z are constant, the above equation may then be expressed as

$$\frac{1}{\rho_f}\frac{\partial (n\,\rho_f)}{\partial t} = K_x \frac{\partial^2 h}{\partial x^2} + K_y \frac{\partial^2 h}{\partial y^2} + K_z \frac{\partial^2 h}{\partial z^2}. \tag{B.16}$$

The specific storage S_s is defined as the change of the specific fluid *volume* per unit change of hydraulic head, i.e.,

$$S_s = \frac{1}{\rho_f}\frac{\partial (n\,\rho_f)}{\partial h} \tag{B.17}$$

which has a unit of m^{-1}.

Assuming that $(n\rho_f)$ is a function of h only, we finally obtain the groundwater flow equation:

$$S_s \frac{\partial h}{\partial t} = K_x \frac{\partial^2 h}{\partial x^2} + K_y \frac{\partial^2 h}{\partial y^2} + K_z \frac{\partial^2 h}{\partial z^2}. \tag{B.18}$$

B.7 Physical Meaning of the Specific Storage

The differential of h in a control volume fixed in space (i.e., $dz = 0$) is

$$dh = \frac{1}{\rho_f\,g} dP \tag{B.19}$$

Then

$$\begin{aligned} S_s = \frac{1}{\rho_f}\frac{\partial (n\,\rho_f)}{\partial h} &= \frac{1}{\rho_f}\left(\rho_f\,g\right)\frac{\partial (n\,\rho_f)}{\partial P} \\ &= \rho_f\,g\,(\beta_n + n\,\beta_w) \end{aligned} \tag{B.20}$$

where $\beta_n \equiv \frac{\partial n}{\partial P}$ and $\beta_w \equiv \frac{1}{\rho_f}\frac{\partial \rho_f}{\partial P}$ are the compressibilities of pores and of the pore fluid, respectively.

The specific storage S_s has a dimension of L^{-1} and is normally much smaller than $10^{-3}\ m^{-1}$. For an aquifer of thickness b, a 'storativity' S may be defined as

$$S = b\,S_s \tag{B.21}$$

which is dimensionless.

B.8 Flow Equation for Isotropic Aquifer

For isotropic aquifer

$$K_x = K_y = K_z = K$$

and the flow equation reduces to

$$S_s = \frac{\partial h}{\partial t} = K\left(\frac{\partial^2 h}{\partial x^2} + \frac{\partial^2 h}{\partial y^2} + \frac{\partial^2 h}{\partial z^2}\right) = K\nabla^2 h. \tag{B.22}$$

Defining the hydraulic diffusivity as

$$D \equiv K/S_s, \tag{B.23}$$

we may express the above equation as

$$\frac{\partial h}{\partial t} = D\nabla^2 h. \tag{B.24}$$

If there is recharge that supplies fluid to the rock/sediments, then

$$\frac{\partial h}{\partial t} = D\nabla^2 h + A/S_s, \tag{B.24a}$$

where A is the rate of fluid supply (in volume per unit rock/sediment volume).

For an aquifer of thickness b (in the z-direction), the flow equation B.24 is often expressed as:

$$S\frac{\partial h}{\partial t} = T\left(\frac{\partial^2 h}{\partial x^2} + \frac{\partial^2 h}{\partial y^2}\right). \tag{B.25}$$

If the flow is in steady-state, the equation further reduces to

$$\nabla^2 h = 0. \tag{B.26}$$

B.9 Calculating Permeability from Tidal Response of Groundwater Level

Hsieh et al. (1987) showed that the transmissivity and the storativity of an aquifer may be calculated from the amplitude response A and phase lag η of groundwater-level oscillations with respect to the tidal forcing by solving the following equations:

$$A = \left(E^2 + F^2\right)^{-1/2} \tag{B.27}$$

$$\eta = -\tan^{-1}(F/E) \tag{B.28}$$

where

$$E \approx 1 - \frac{\omega\, r_c^2}{2T}\, Kei\,(\alpha),\ F \approx \frac{\omega\, r_c^2}{2T}\, Ker\,(\alpha), \text{ and } \alpha = \left(\frac{\omega S_s}{T}\right)^{\frac{1}{2}} r_w$$

The amplitude response A is the ratio between the amplitude of the water level fluctuation and that of the pressure head fluctuation, T is transmissivity, S_s the specific storage, ω the frequency of the oscillation, r_w and r_c the radius of the well and the inner radius of well casing, respectively, *Ker* and *Kei* the zeroth order Kelvin functions. Since A, η, ω, r_w and r_c are all measured values, T and S_s may be calculated by solving the above equations. Finally, the permeability k may be calculated from T using the relation (Appendix B) $k = \mu T / b \rho_f g$, where μ and ρ_f are, respectively, the viscosity and density of water, b the thickness of the aquifer and g the gravitational acceleration.

B.10 Equation Derivations

1) *Derivation of Eq. (2.6)*
In 1D, Eq. (B.24a) is simplified to:

$$S_s \frac{\partial h}{\partial t} = K \frac{\partial^2 h}{\partial x^2} + A.$$

We solve this equation under the boundary conditions of no-flow at $x = 0$ (i.e., a local water divide) and $h = 0$ at $x = L$. If A is a function of x only, the solution is given by (Carslaw and Jaeger, 1959, p. 132, Eq. 10):

$$h(x,t) = \frac{4L}{\pi^2 K} \sum_{n=1}^{\infty} \frac{1}{n^2} \left[1 - \exp\left[-\frac{D n^2 \pi^2 t}{4L^2}\right]\right] \times \cos\frac{n\pi x}{2L} \int_{-L}^{L} A(x) \cos\frac{n\pi x'}{2L} dx'$$

Since the source (or sink) for the present study is a function of both x and t, we may apply the Duhamel's principle (Carslaw and Jaeger, 1959, p. 32, Eq. 20) to the above equation and obtain

$$h\,(x,t) = \frac{1}{L S_s} \sum_{n=1}^{\infty} \cos\frac{n\pi x}{2L} \int_{-L}^{L} \int_0^t \exp\left[-\frac{D n^2 \pi^2 (t - \lambda)}{4L^2}\right] \times A(x,\lambda) \cos\frac{n\pi x'}{2L}\, d\lambda\, dx'$$

Furthermore, since the time duration for the coseismic release of water is very much shorter than the time duration for the postseismic evolution of groundwater level, we may consider the coseismic release of water to be instantaneous; i.e.,

$$A\,(x,t) = A_0\,(x)\,\delta\,(t = 0)$$

where $A_o(x)$ is the spatial distribution of the earthquake-induced fluid source and $\delta(t = 0)$ is the delta function that equals 1 when $t = 0$ and equals zero when $t > 0$. The above equation is then reduced to

$$h(x,t) = \frac{1}{L S_s} \sum_{n=1}^{\infty} \cos \frac{n \pi x}{2L} \exp\left[-\frac{D n^2 \pi^2 t}{4 L^2}\right] \int_{-L}^{L} H_o(x') \cos \frac{n\pi x'}{2L} dx'$$

where

$$H_o(x) = \int_0^t A_o(x)\,\delta(t)\,dt$$

The function $A_o(x)$ has a unit of volume per volume per time, while the function $H_o(x)$ has a unit of volume per volume. If we further simplify the problem by assuming $H_o(x) = H_o$ for $x \leq L'$ and 0 for $x > L'$ (Fig. 4.5c), the above expression for $h(x,t)$ reduces to (4.6).

$$h(x,t) = \frac{4H_o}{\pi S_s} \sum_{n=1}^{\infty} \frac{1}{n} \sin \frac{n\pi L'}{2L} \cos \frac{n\pi x}{2L} \exp\left[-n^2 \frac{t}{\tau}\right].$$

2) *Derivation of Eq. (5.7)*
During earthquakes the entire aquifer in the alluvial fan, i.e., $L' = L$, may consolidate and supply an extra amount of water H_o per unit volume. In this case, only the odd terms in the above summation are meaningful. Thus the above equation becomes

$$h(x,t) = \frac{4 H_o}{\pi S_s} \sum_{r=1}^{\infty} \frac{(-1)^{r+1}}{(2r-1)} \cos \frac{(2r-1)\pi x}{2L} \exp\left[-(2r-1)^2 \frac{t}{\tau}\right],$$

which is Eq. (5.7).

References

Athy, L.F., 1930, Density, porosity, and compaction of sedimentary rocks, *Am. Assoc. Petroleum Geologists Bull., 14*, 1–24.
Carslaw, H.S., and J.C. Jaeger, 1959, *Conduction of Heat in Solids*, Oxford: Clarendon Press.
Deming, D., 2002, *Introduction to Hydrogeology*, New York: McGraw Hill.
Dominico, P.A., and F.W. Schwarz, 1998, *Physical and Chemical Hydrogeology*, 2nd ed., New York: John Wiley.
Freeze, R.A., and J.A. Cherry, 1979, *Groundwater*, Englewood Cliffs: Prentice-Hall.
Hsieh, P., J. Bredehoeft, and J. Farr, 1987, Determination of aquifer permeability from earthtide analysis, *Water Resour. Res., 23*, 1824–1832.
Ingebritsen, S.E., W.E. Sanford, and C.E. Neuzil, 2006, *Groundwater in Geologic Processes*, 2nd ed., New York: Cambridge University Press.

Manning, C.E., and S.E. Ingebritsen, 1999, Permeability of the continental crust: Constraints from heat flow models and metamorphic systems, *Rev. Geophys.*, *37*, 127–150.
Saar, M.O., and M. Manga, 2004, Depth dependence of permeability in the Oregon Cascades inferred from hydrogeologic, thermal, seismic, and magmatic modeling constraints, *J. Geophys. Res.*, *109*, B04204, doi:10.1029/2003JB002855.

Appendix C: Groundwater Transport

In this appendix we provide the basic equations for heat and solute transport. We then derive their respective Peclet numbers as referred to in Chap. 4.

C.1 Governing Equations for Heat Transport

Heat *conduction* is governed by Fourier's law

$$q_h = -K_h \frac{dT}{dx} \tag{C.1}$$

where T is temperature, and K_h is thermal conductivity of the solid. Unlike permeability, K_h varies by less than a factor of five for rocks and sediments. Clay, one of the least conductive material, has $K_h = 1\ \mathrm{Wm^{-1}K^{-1}}$, while granite, a good conductor, has $K_h = 3\ \mathrm{Wm^{-1}K^{-1}}$. For saturated porous media, heat conductivity may be estimated from

$$K_h \approx nK_f + (l - n)\,K_r \tag{C.2}$$

where K_f and K_r are, respectively, the heat conductivity of the pore fluid and the solid rock. At 25°C, K_f is about 6 $\mathrm{Wm^{-1}K^{-1}}$ and can dominate the thermal conductivity of saturated sediments and porous rocks.

Combining Fourier's law with the law of conservation of energy, we obtain a differential equation for the conductive heat transport,

$$\left[n\rho_f c_f + (1 - n)\,\rho_r c_r\right] \frac{\partial T}{\partial t} = K_h \nabla^2 T. \tag{C.3}$$

where c_f and c_r are the specific heat of the pore fluid and the solid rock, respectively.

The amount of heat carried by the flowing fluid, i.e., by *advection*, is

$$\rho_f\, c_f\, \nabla T \cdot \boldsymbol{q} \tag{C.4}$$

where $\boldsymbol{q} = K\nabla h$ is Darcy's velocity. Accounting for advection and heat production, the heat transport equation becomes

$$\left[n\rho_f c_f + (1 - n)\,\rho_r c_r\right] \frac{\partial T}{\partial t} = K_h \nabla^2 T - \rho_f c_f \boldsymbol{q} \cdot \nabla T + Q_h \tag{C.5}$$

where Q_h is the heat produced (+) or absorbed (−) per unit volume. Because of the presence of $\boldsymbol{q}$ in the equation, the solution of the heat transport is 'coupled' to the groundwater flow.

C.2 Relative Significance of Advective Versus Conductive Heat Transport

From Eq. (C.5), we see that the magnitude of the conductive heat transport is of the order of $K_h\ \Delta T/L^2$ and the magnitude of the advective heat transport is of the order of $\rho_f c_f q \Delta T/L$, where L is the linear dimension of the studied object. Hence the relative significance of the advective versus the conductive heat transport is given by the ratio

$$\mathrm{Pe}_h = \frac{\rho_f c_f q \Delta T/L}{K_h \Delta T/L^2} = \frac{\rho_f c_f q L}{K_h}. \tag{C.6}$$

This dimensionless ratio is known as the Peclet number for heat transport. If Pe_h is greater than 1, the advective heat transport is more important than the conductive heat transport, and vice versus.

C.3 Governing Equations for Solute Transport

Let C be the concentration of a chemical component in water. If there is a gradient in C, *diffusion* will occur according to Fick's law

$$q_d = -D_w \frac{dC}{dx} \tag{C.7}$$

where D_w is the coefficient of molecular diffusion in water.

In porous media, the diffusion of chemical components is impeded by the tortuous paths through which the water flows, in addition to the limited pore space. The diffusion coefficient in porous media is related to D_w by

$$D_m \approx \frac{n_e}{\tau} D_w \tag{C.8}$$

where n_e is the effective porosity (i.e., pore space in which chemical components move unimpeded with flow) and τ is tortuosity (i.e., the ratio between the path length through which a molecule actually moves from one point to another and the straight-line distance between the two points). Typical values for the diffusion coefficients for geologic media range from 10^{-11} to 10^{-10} m^2/s. In geologic media, Fick law for diffusion is modified as

$$q_d = -D_m \frac{dC}{dx}. \tag{C.9}$$

By combining Fick's relation for geologic media with the conservation law for the chemical mass, we obtain the transport equation for diffusion:

$$n_e \frac{\partial C}{\partial t} = D_m \nabla^2 C. \quad \text{(C.10)}$$

C can also change by *advection* according to

$$\frac{\partial C}{\partial t} = -v_x \frac{\partial C}{\partial x}, \quad \text{(C.11)}$$

where v_x is the average linear velocity in the x-direction. The linear velocity $\boldsymbol{v}$ is related to the Darcy's velocity $\boldsymbol{q}$ by

$$\boldsymbol{v} = \boldsymbol{q}/n_e \quad \text{(C.12)}$$

Mechanical *dispersion* is caused by the 3-D microscopic heterogeneities in the porous media. In an isotropic media, the dispersivity may be expressed by two components: α_L, the dispersivity along the direction of groundwater flow, and α_T, the dispersivity transverse to the direction of groundwater flow. The sum of the mechanical and diffusive transport is the 'hydrodynamic' dispersion, given by

$$\begin{aligned} D_L &= \alpha_L v + D'_m \\ D_T &= \alpha_T v + D'_m \end{aligned} \quad \text{(C.13)}$$

where D_L and D_T are the longitudinal and transverse dispersion coefficients and $D_m{}' = D_m/n_e$.

We can now rewrite the equation for solute transport to include both the advective and the dispersive effect, as well as the source (or sink) for the solute:

$$\frac{\partial \left(n_e \rho_f C\right)}{\partial t} = \nabla \cdot \left(n_e \rho_f \underline{D} \nabla C\right) - \nabla \cdot \left(n_e \rho_f v C\right) + Q_s \quad \text{(C.14)}$$

where $\underline{D}$ is the hydrodynamic dispersion tensor, and Q_s is a source (+) or sink (–) of solute per unit volume. Because of the presence of v in the above equation, solute transport is 'coupled' to groundwater flow.

For systems with constant fluid density and constant effective porosity that is equal to the total porosity, the above transport equation reduces to

$$\frac{\partial C}{\partial t} = \nabla \cdot \left(\underline{D} \nabla C\right) - \nabla \cdot (v C) + \frac{Q_s}{n \rho_f} \quad \text{(C.15)}$$

C.4 Relative Significance of Advective Versus Diffusive Solute Transport

In Eq. (C.15), we see that the magnitude of the diffusive solute transport is of the order of $D\Delta C/L^2$ and the magnitude of the advective heat transport is of the order of $v\Delta C/L$, where L is the linear dimension of the studied object. Hence the relative significance of the advective versus the diffusive solute transport is given by the ratio

$$\mathrm{Pe}_c = \frac{v\,\Delta C/L}{D\,\Delta C/L^2} = \frac{v\,L}{D} \tag{C.16}$$

which is known as the Peclet number for solute transport. If Pe_c is greater than 1, advective transport is more important than the diffusive transport, and vice versa.

Appendix D: Hydromechanical Coupling

D.1 Introduction

Under applied loading, porous rock/sediments deform by changing the pore space as well as by straining the solid matrix. If the pore space is filled with fluid, the pressure in the pore fluid may change in response to the change in pore volume. If a gradient is generated in the pore pressure, fluid flow will take place along the hydraulic gradient (Appendix B). At the same time, a change in pore pressure may affect the properties of the porous medium, such as its permeability, strength, and stress-strain relation. Thus there is a hydromechanical coupling between the pore fluid and the solid matrix.

Studies of the hydromechanical coupling between pore fluid and the solid matrix have followed a poroelastic approach. Biot (1941) laid the theoretical foundation for poroelasticity which describes the hydromechanical coupling for linear elastic media. This approach has been applied to the study of a variety of problems such as the role of pore pressure on seismicity, pore pressure generation under tectonic stress, and pore pressure change in response to erosional unloading.

However, rocks and sediments remain elastic only if the strain level is small, i.e., $<10^{-4}$ (Fig. D.1). Above this threshold, the deformation of most rock/sediments becomes non-linear, irreversible and even time dependent. The concept of effective stress may continue to hold, but the assumption of linear elasticity breaks down.

In the following we first introduce the effective stress principle and derive the equations for poroelasticity and hydromechanical coupling that are noted in Chaps. 8 and 9 in the book. Unlike the linear elastic media where a theoretical foundation (poroelasticity) is available for hydromechanical coupling, most studies on the hydromechanical coupling for non-elastic deformation of fluid-saturated porous

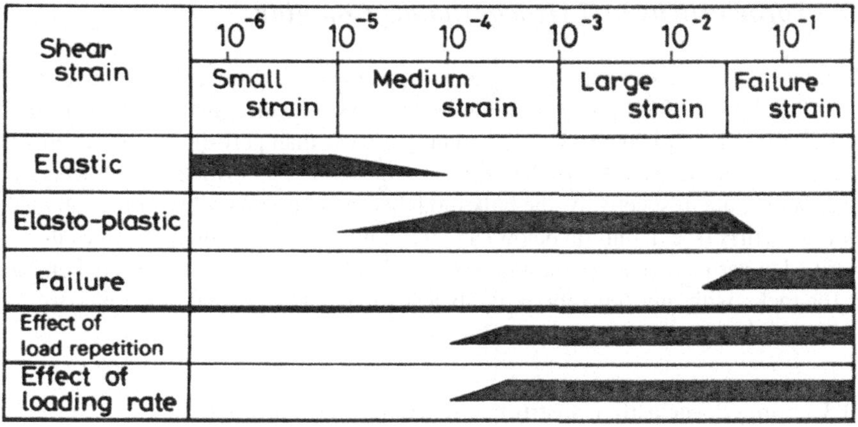

Fig. D.1 Schematic drawing to show the ranges of strains for different mechanical responses of rocks and sediments to applied load. Note that the threshold strain for failure decreases drastically under repeated or long duration loading (From Ishihara, 1966)

media have been empirical, and we show some experimental results for the non-elastic deformation of sediments under cyclic loading, which is relevant to Chaps. 3, 5 and 6 in the book.

D.2 Effective Stress Principle

Experiments show that the effect of pore pressure on the properties of the solid matrix manifests frequently through an 'effective' stress principle. Terzaghi (1923) first proposed the concept of effective stress to explain observed time-dependent subsidence of land under surface load:

$$\sigma'_{zz} = \sigma_{zz} - \alpha P \tag{D.1}$$

where σ'_{zz} is the effective stress in the vertical direction, σ'_{zz} is the vertical stress, P is the pore pressure in the porous rock or sediments, and α is an empirical parameter (the Biot-Willis coefficient) to be determined. In words, the pore pressure generated by the surface load acts to reduce the effective vertical stress by an amount αP. As pore pressure dissipates, the effective stress increases, causing time-dependent consolidation of rock/sediments and thus land subsidence.

Although Terzaghi's expression for the effective stress is specific for the vertical loading, the concept of effective stress is not restricted in the vertical direction. The general expression for the effective stress is

$$\sigma'_{ij} = \sigma_{ij} - \alpha P \delta_{ij} . \tag{D.2}$$

D.3 Poroelasticity and Hydrodynamic Coupling

What is α? Formerly, α was thought to be equal to porosity since the effective stress is reduced by the fluid pressure in the pores. However, the value of α, determined experimentally, turns out to be substantially greater than porosity. It can be proved through the poroelastic theory that $\alpha = 1 - \frac{K}{K_s} = 1 - \frac{c_s}{c}$ (Wang, 2000), where K, c, K_s and c_s are, respectively, the bulk modulus and compressibility of the 'drained' porous matrix (see definition below) and the solid grains. This theoretical result was verified experimentally by Nur and Byerlee (1971). Since c_s is always smaller than c for rocks with any porosity, α is always smaller than 1; and since $c_s \ll c$ for unconsolidated sediments, α is nearly 1 for these materials.

D.3.1 Some Poroelastic Constitutive Relations and Parameters

Following is a brief introduction to some poroelastic constitutive relations and parameters that were referred to in Chaps. 8 and 9. For an excellent review of the topic, see Wang (2000). Consider a saturated porous rock in isothermal condition. The volumetric strain ε and fluid content f in the rock are thus functions of the isotropic stress σ and pore pressure P only; thus

$$d\varepsilon = \left(\frac{\partial \varepsilon}{\partial \sigma}\right)_P d\sigma + \left(\frac{\partial \varepsilon}{\partial P}\right)_\sigma dP \tag{D.3}$$

$$df = \left(\frac{\partial f}{\partial \sigma}\right)_P d\sigma + \left(\frac{\partial f}{\partial P}\right)_\sigma dP \tag{D.4}$$

Biot (1941) defined four parameters for the partial derivatives in the above equations:

$$\frac{1}{K} = \left(\frac{\partial \varepsilon}{\partial \sigma}\right)_P \tag{D.5}$$

$$\frac{1}{H} = \left(\frac{\partial \varepsilon}{\partial P}\right)_\sigma \tag{D.6}$$

$$\frac{1}{H_1} = \left(\frac{\partial f}{\partial \sigma}\right)_P \tag{D.7}$$

and

$$\frac{1}{R} = \left(\frac{\partial f}{\partial P}\right)_\sigma . \tag{D.8}$$

It can be proved that $H = H_1$ (Wang, 2000); thus there are only three independent parameters.

Two additional parameters, the Skempton's coefficient and the Biot-Willis coefficient, are often encountered, as we did in Chaps. 8 and 9. The Skempton's coefficient is defined as the ratio between the change in pore pressure and the change in the isotropic stress at constant fluid content, i.e.,

$$B = -\left(\frac{\partial P}{\partial \sigma}\right)_f. \tag{D.9}$$

Given

$$-\left(\frac{\partial P}{\partial \sigma}\right)_f = \left(\frac{\partial f}{\partial \sigma}\right)_P \Big/ \left(\frac{\partial f}{\partial P}\right)_\sigma$$

we have

$$B = \frac{R}{H}. \tag{D.10}$$

The Biot-Willis coefficient is defined as the ratio between the change in pore pressure and the change in the isotropic stress at constant volumetric strain, i.e.,

$$\alpha = -\left(\frac{\partial P}{\partial \sigma}\right)_\varepsilon. \tag{D.11}$$

Following the same procedure as above, we have

$$\alpha = \frac{K}{H}. \tag{D.12}$$

The parameter K, as defined in (D.5), is the bulk modulus at constant pore pressure and is sometime referred to as the 'drained' bulk modulus because it is measured in a condition where the pore fluid in the rock/sediment specimen is hydraulically open to an external reservoir of constant pressure P; thus fluid is free to drain in or out of the rock/sediment specimen. We may similarly define an 'undrained' bulk modulus which is measured in a condition where the fluid in the specimen is completely confined in the rock/sediments; thus the fluid content of the specimen is constant; i.e.,

$$\frac{1}{K_u} = \left(\frac{\partial \varepsilon}{\partial \sigma}\right)_f \tag{D.13}$$

The 'undrained' bulk modulus and the 'drained' bulk modulus may be shown to be related by the following relation (Wang, 2000):

$$K_u = \frac{K}{1 - \alpha B}. \tag{D.14}$$

Given the above relations, we may invert Eqs. (D.3) and (D.4), respectively, to obtain the following expressions:

$$d\sigma = K_u\, d\varepsilon - K_u B\, df, \tag{D.15}$$

and

$$dP = -K_u B\, d\varepsilon + \frac{K_u B}{\alpha}\, df. \tag{D.16}$$

Under undrained conditions ($df = 0$), the change in the pore pressure in rock/sediments is thus proportional to the change of the volumetric strain by the following relation

$$dP = -K_u B\, d\varepsilon\,, \tag{D.17}$$

as stated in Chap. 9.

D.3.2 General Constitutive Relations for Poroelastic Media

$$\begin{aligned}
\sigma_{xx} - \alpha P &= 2G\left(\varepsilon_{xx} + \frac{\nu}{1-2\nu}\varepsilon_{kk}\right)\\
\sigma_{yy} - \alpha P &= 2G\left(\varepsilon_{yy} + \frac{\nu}{1-2\nu}\varepsilon_{kk}\right)\\
\sigma_{zz} - \alpha P &= 2G\left(\varepsilon_{zz} + \frac{\nu}{1-2\nu}\varepsilon_{kk}\right)\\
\sigma_{xy} &= 2G\,\varepsilon_{xy}\\
\sigma_{yz} &= 2G\,\varepsilon_{yz}\\
\sigma_{xz} &= 2G\,\varepsilon_{xz}
\end{aligned} \tag{D.18}$$

where G is the rigidity modulus and ν is the Poisson's ratio. In tensor notation, these relations may be expressed as

$$\sigma_{ij} - \alpha P \delta_{ij} = 2G\left(\varepsilon_{ij} + \frac{\nu}{1-\nu}\varepsilon_{kk}\delta_{ij}\right), \tag{D.19}$$

The above expression may be converted to another form (Wang, 2000)

$$\varepsilon_{ij} = \frac{1}{2G}\left(\sigma_{ij} - \frac{\nu_u}{1+\nu_u}\sigma_{kk}\delta_{ij} + \frac{2GB}{3} f\,\delta_{ij}\right), \tag{D.20}$$

where ν_u is the 'undrained' Poisson's ratio and is related to ν by

$$\nu_u = \frac{3\nu + \alpha B(1-2\nu)}{3 - \alpha B(1-2\nu)} \tag{D.21}$$

D.3.3 Groundwater Flow Equations for Poroelastic Media

Recall that in formulating the equation for groundwater flow (Appendix B), we assumed that the hydraulic head h at a given elevation is depends only on the pore pressure P. In poroelastic media, h is a function of P as well as stress σ. Let $f \equiv n\rho_f$, we have

$$\frac{\partial f}{\partial t} = \left(\frac{\partial f}{\partial P}\right)_\sigma \frac{\partial P}{\partial t} + \left(\frac{\partial f}{\partial \sigma}\right)_P \frac{\partial \sigma}{\partial t}. \tag{D.22}$$

Defining the specific storage at constant stress as

$$s_\sigma \equiv \left(\frac{\partial f}{\partial P}\right)_\sigma, \tag{D.23}$$

and the ‘drained’ bulk modulus as

$$K \equiv \left(\frac{\partial \sigma}{\partial \varepsilon}\right)_P. \tag{D.24}$$

We can show (Wang, 2000)

$$s_\sigma = \frac{\alpha}{KB}, \tag{D.25}$$

and

$$\left(\frac{\partial f}{\partial \sigma}\right)_P = \frac{\alpha}{K}. \tag{D.26}$$

Following the same procedure as in Appendix B, we obtain the groundwater flow equation in poroelastic media:

$$s_\sigma \left(\frac{\partial P}{\partial t} + \frac{B}{3}\frac{\partial \sigma_{kk}}{\partial t}\right) = \frac{k}{\mu}\nabla^2\left(P + \rho_f g z\right). \tag{D.27}$$

The presence of the term $\partial\sigma_{kk}/\partial t$ on the left-hand side of this equation makes it different from the flow equation developed in Appendix B.

D.4 Non-elastic Deformation

The study of deformation in nature is complicated by several factors. On short time scales (seconds) and small stresses, geologic media may deform as elastic material. On longer time scales and large stresses, however, they may deform by flow or fracture. The relative motion between Earth’s tectonic plates causes shearing near the plate boundaries; the accumulated shear strain could be large – far exceeding the threshold for linear elasticity (Fig. D.1). Rocks and sediments under such conditions may suddenly fail by faulting (earthquake) if the confining pressure and temperature are relatively low (upper crustal condition), or they may yield by flow if the confining pressure and temperature are high. Cyclic loading, such as that associated with seismic waves, may also induce non-elastic deformation and failure even if the amplitude of stress is small (Fig. D.1). In the following, we show some results of cyclic experiments that are relevant to the discussions in Chaps. 3, 5 and 6.

D.5 Deformation Under Cyclic Loading

Since early 1960s, a great number of laboratory experiments has been carried out by geotechnical engineers to study the consolidation of sediments under cyclic loading.

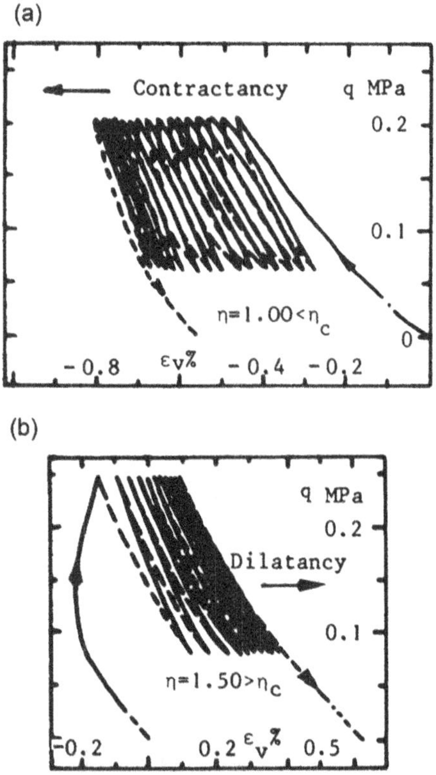

Fig. D.2 Diagrams of deviatoric stress versus volumetric strain for two sand samples of the same constitution. (**a**) Shearing at a maximum deviatoric stress '*q*' of 0.2 MPa. Note that the volumetric strain decreases with increasing number of stress cycles and the sample contracts under cyclic shearing. (**b**) Shearing at a maximum deviatoric stress of 0.3 MPa. Note that the volumetric strain increases with increasing number of stress cycles and the sample dilates under cyclic shearing. A 'critical state' may thus exist between 0.2 and 0.3 MPa deviatoric stress, where no contraction or dilatancy occurs (From Luong, 1980)

Figure D.2 shows the deformation of porous sediments under drained condition (Luong, 1980). At shear stresses below a characteristic threshold (Fig. D.2a), cyclic shearing causes the volume of the sheared sample to decrease, while at shear stress above the threshold (Fig. D.2b), cyclic shearing causes the volume of the sheared sample to increase. The latter condition may occur in the immediate vicinity of a

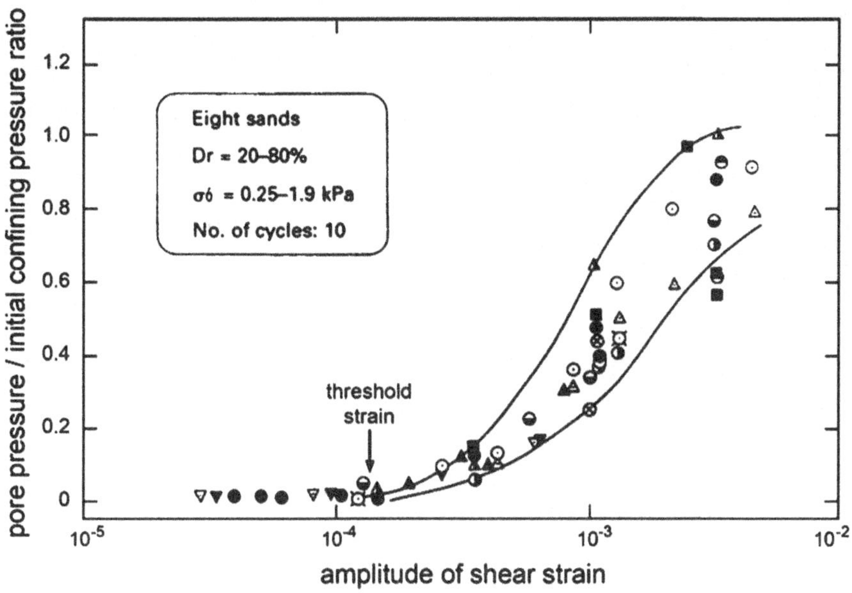

Fig. D.3 Experimental results for pore-pressure generation in eight different sands with dry density from 20 to 80% of the mineral density, under initial confining pressure from 0.25 to 1.9 KPa, and a uniform 10 cycles of loading. Note that, for sands with such diverse densities and confining pressures, pore-pressure buildup occurs at a threshold strain of 10^{-4} (From Vucetic, 1994)

ruptured fault. Since most observations are at some distance away from the ruptured fault, the case in Fig. D.2a normally applies. Thus, if the sheared rock/sediments are saturated, we normally expect that pore pressure increases with cyclic shearing. Using a wide variety of saturated sediments under a wide range of confining pressures, Dobry et al. (1982), Vucetic (1994) and Hsu and Vucetic (2004) showed that pore pressure begins to buildup when sediments are sheared above a strain amplitude of 10^{-4} (Fig. D.3), which exemplify a threshold strain amplitude for the initiation of sediment consolidation.

Figure D.4 shows how deformation and pore pressure increased in a sand specimen under cyclic shearing at constant stress amplitude of ±50 KPa. Note that the axial strain was small during the first six cycles, but increased significantly afterwards. The large stains at cycles 8, 9 and 10 show that the sample was failing. Pore pressure started to build up as soon as the cyclic shear began. The sinusoidal pore-pressure change was in phase with the applied load until the last few cycles where the pore pressure dropped near the peak shear stress, suggesting that the sample dilated during these cycles at the peak shear stress. When pore pressure becomes equal to the overburden (or confining) pressure, sediments lose shear strength and become fluid-like. The experiment shows that in cyclic loading, failure is not due to the peak stress but to the accumulated work of many cycles. The result in Fig. D.5

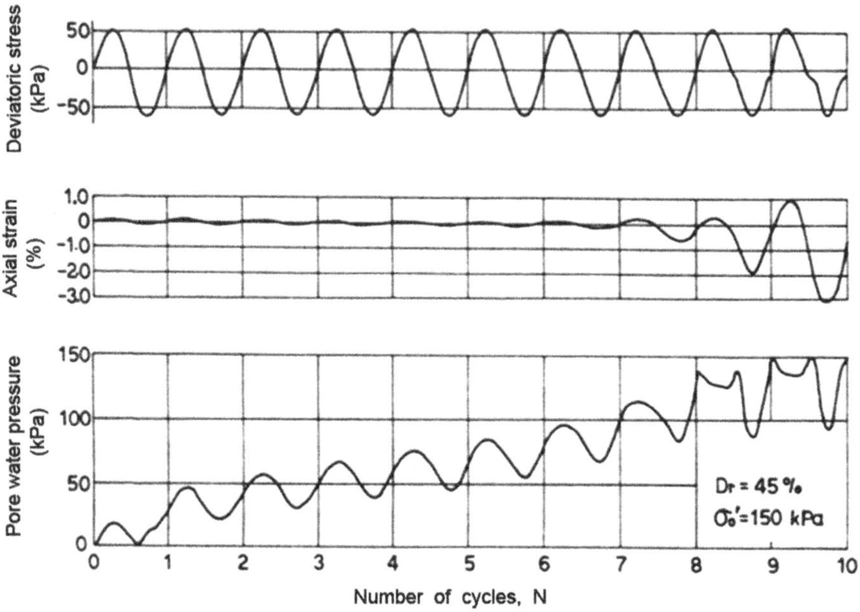

Fig. D.4 Experimental results of pore-pressure generation and axial deformation in a saturated sand specimen subjected to cyclic shearing at constant amplitude of ±50 KPa. The sediment specimen had an initial dry density of 45% and was subjected to a confining pressure of 150 KPa. Note that the deformation was small during the first six cycles, but increased significantly afterwards. The large deformation at cycle 8, 9 and 10 suggests that the sample was failing. Pore-pressure started to build up at the beginning of shearing and showed sinusoidal changes in phase with the applied shearing until the last few cycles, where pore pressure dropped near the peak shearing, suggesting that the sample may have dilated at the peak shearing (From Seed and Lee, 1966)

bears further evidence to this statement. These changes are complimentary to those shown in Fig. 2.5 in a constant strain experiment (Sect. 2.).

The results of the above experiment may be plotted in a stress versus strain diagram, as shown in Fig. D.5a. The loss in rigidity, and thus the occurrence of liquefaction, is marked by the drastic increase in shear strain at the end of the experiment. The stress-strain loops in such diagrams may be integrated to estimate the dissipated energy required to initiate pore-pressure buildup and liquefaction in cyclic shearing, as discussed in Chaps. 2 and 5. Notice the great difference between the response of the loose sand (Fig. D.5a) and that of the dense sand (Fig. D.5b). Many more stress cycles (or greater shear stresses) are needed to liquefy the dense sand.

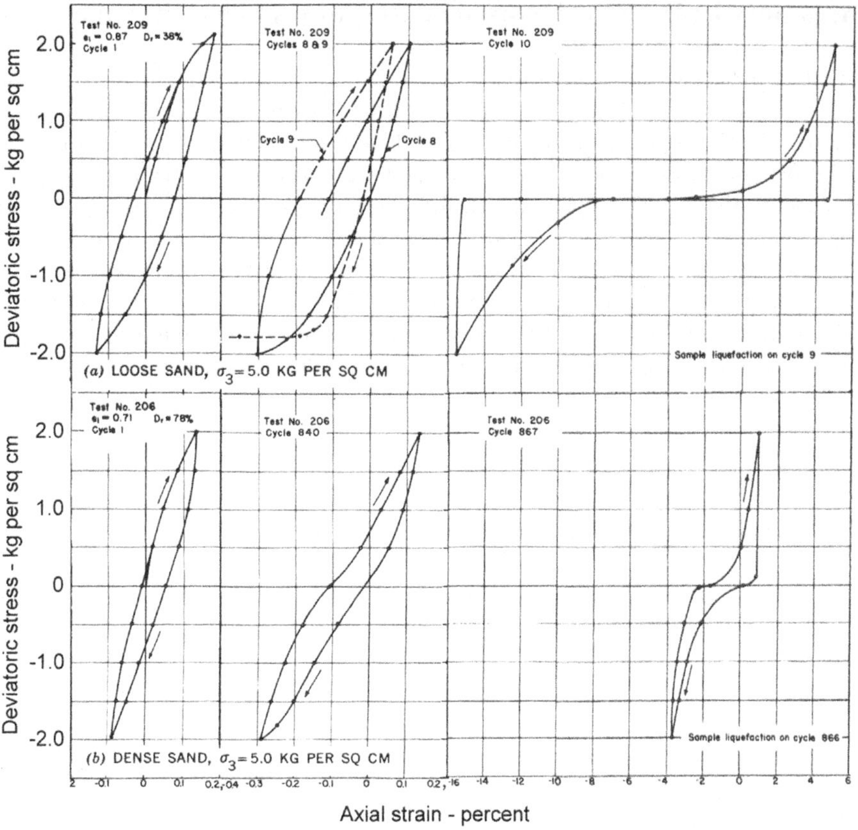

Fig. D.5 Stress-strain diagram for sand specimen under cyclic shearing at constant stress amplitude. Note that the hysteresis loop, moderate at the earlier cycles, becomes extreme at the end of the experiment, indicating that the sample had liquefied (From Seed and Lee, 1966)

References

Ishihara, K., 1996, *Soil Behavior in Earthquake Geotechnics*, 350 pp, Oxford: Clarendon Press.

Luong, M.P., 1980, Stress-strain aspects of cohesionless soils under cyclic and transient loading. In: G.N. Pande, and O.C. Zienkiewicz (eds.), *Proc. Intern. Symp. Soils under Cyclic and Transient Loading*, pp. 315–324, Rotterdam, Netherlands: A.A. Balkema.

Seed, H.B., and K.L. Lee, 1966, Liquefaction of saturated sands during cyclic loading, *J. Soil Mech. Found. Div., 92*, 105–134.

Segall, P., J.R. Grasso, and A. Mossop, 2004, Poroelastic stressing and induced seismicity near the Lacq gas field, southwestern France, *J. Geophys. Res., 99*, 15423–15438.

Vucetic, M., 1994, Cyclic threshold of shear strains in soils, *J. Geotech. Eng., 120*, 2208–2228.

Wang, H.F., 2000, *Theory of Linear Poroelasticity*, 287 pp, Princeton: Princeton University Press.

Appendix E: Data for Hydrologic Responses to Earthquakes

In this appendix we tabulate the data used to construct the magnitude-distance diagrams in this book (Figs. 2.9, 3.6, 5.6, 6.2 and 10.1) for ready access by the readers. Readers are nevertheless encouraged to read the listed references before using this data. We note that the tables do not generally share the same format because, first, the data in the different tables are of different types, and second, the time period represented by some datasets, such as liquefaction, range from the twelfth to the twenty-first centuries, and the available information changes with time. We also note that, because a large amount of historical data is included in some datasets, for which the distinction among different scales for earthquake magnitude is not available, we choose to neglect the distinction among the different magnitude scales. The high-frequency data used in Sect. 5.5, however, cannot be included because of its excessive volume.

E.1 Stream and Spring Responses

Stream/spring	Earthquake	Magnitude	Distance (km)	References
Alum Rock, CA	Alum Rock 18 Oct 1989	5.0	5	King et al. (1994)
Alum Rock, CA	Alum Rock 30 Oct 2007	5.2	4	Rowland et al. (2008)
Alum Rock, CA	Alum Rock 13 Jun 1988	5.3	8	King et al. (1994)
Alum Rock, CA	Mount Lewis 31 Mar 1986	5.7	15	King et al. (1994)
Mattole River, CA	Honeydew 17 Aug 1991	6.1	7	McPherson and Dengler (1992)
Alum Rock, CA	Morgan Hill 24 Apr 1984	6.2	18	King et al. (1994)
Salinas River, CA	San Simeon 22 Dec 2003	6.5	38	Wang et al. (2004)
Lopez Creek, CA	San Simeon 22 Dec 2003	6.5	72	Wang et al. (2004)
Sespe Creek, CA	San Fernando 9 Feb 1971	6.6	47	Manga et al. (2003)
Sespe Creek, CA	Northridge 17 Jan 1994	6.7	44	Manga et al. (2003)
Santa Paula Creek, CA	Northridge 17 Jan 1994	6.7	55	Quilty et al. (1995)
Various, Washington	Nisqually 28 Feb 2001	6.8	13, 16, 18, 20, 23, 24, 26, 29, 30, 34, 40, 42, 44, 45, 46, 47, 48, 49, 50, 51, 52, 53, 54, 55, 58, 60, 63, 65, 66, 69, 79, 80, 85, 87, 88, 90, 92, 97, 109, 114	Montgomery et al. (2003)
Big Lost River, Mackay	Borah Peak 28 Oct 1983	7.0	12	Muir-Wood and King (1993)
Warm Springs Cr	Borah Peak 28 Oct 1983	7.0	13	Muir-Wood and King (1993)
Little Lost River	Borah Peak 28 Oct 1983	7.0	40	Muir-Wood and King (1993)
Thomson, Bruno, and Squaw Cr	Borah Peak 28 Oct 1983	7.0	51	Muir-Wood and King (1993)
Little Wood River	Borah Peak 28 Oct 1983	7.0	64	Muir-Wood and King (1993)
Big Wood River	Borah Peak 28 Oct 1983	7.0	68	Muir-Wood and King (1993)
Salmon River	Borah Peak 28 Oct 1983	7.0	78	Muir-Wood and King (1993)
S Fork, Boise River	Borah Peak 28 Oct 1983	7.0	100	Muir-Wood and King (1993)
Soquel Cr	Loma Prieta 18 Oct 1989	7.1	4	Muir-Wood and King (1993)
Carbonera Cr	Loma Prieta 18 Oct 1989	7.1	6	Muir-Wood and King (1993)
Bean Cr	Loma Prieta 18 Oct 1989	7.1	8	Muir-Wood and King (1993)

(continued)

Stream/spring	Earthquake	Magnitude	Distance (km)	References
Corralitos Cr	Loma Prieta 18 Oct 1989	7.1	10	Muir-Wood and King (1993)
Zayante Cr	Loma Prieta 18 Oct 1989	7.1	12	Muir-Wood and King (1993)
San Lorenzo River	Loma Prieta 18 Oct 1989	7.1	13	Muir-Wood and King (1993)
Bear Cr	Loma Prieta 18 Oct 1989	7.1	19	Muir-Wood and King (1993)
Saratoga Cr	Loma Prieta 18 Oct 1989	7.1	25	Muir-Wood and King (1993)
Boulder Cr	Loma Prieta 18 Oct 1989	7.1	27	Muir-Wood and King (1993)
San Lorenzo Cr	Loma Prieta 18 Oct 1989	7.1	28	Muir-Wood and King (1993)
Pescadero Cr	Loma Prieta 18 Oct 1989	7.1	35	Muir-Wood and King (1993)
Brown House Sp	Loma Prieta 18 Oct 1989	7.1	36	Briggs (1991)
Waddell Cr	Loma Prieta 18 Oct 1989	7.1	39	Briggs (1991)
Alum Rock, CA	Loma Prieta 18 Oct 1989	7.1	40	King et al. (1994)
San Gregario Cr	Loma Prieta 18 Oct 1989	7.1	46	Muir-Wood and King (1993)
San Francisquito Cr	Loma Prieta 18 Oct 1989	7.1	54	Muir-Wood and King (1993)
Pilarcitos Cr	Loma Prieta 18 Oct 1989	7.1	61	Muir-Wood and King (1993)
	Loma Prieta 18 Oct 1989	7.1	86	Rojstaczer et al. (1995) as reported in Montgomery et al. (2003)
Various, Awaji Island	Kobe 17 Jan 1995	7.2	3, 4, 5, 6, 7, 8, 10, 11, 13, 18, 19, 20, 21, 22, 23, 25, 26, 27, 29, 34, 36, 38, 40, 43, 44, 45, 47, 49, 50	Sato et al. (2000)
Gardiner River	Hebgen Lakek 18 Aug 1959	7.3	37	Muir-Wood and King (1993)
Madison River	Hebgen Lake 18 Aug 1959	7.3	41	Muir-Wood and King (1993)
Gallatin River	Hebgen Lake 18 Aug 1959	7.3	46	Muir-Wood and King (1993)
Red Rock River	Hebgen Lake 18 Aug 1959	7.3	57	Muir-Wood and King (1993)
Henry's Fork	Hebgen Lake 18 Aug 1959	7.3	60	Muir-Wood and King (1993)
Ruby River	Hebgen Lake 18 Aug 1959	7.3	62	Muir-Wood and King (1993)
Sespe Creek, CA	Landers 28 Jun 1992	7.3	225	Manga et al. (2003)
Millard Creek, CA	Landers 28 Jun 1992	7.3	47	Roeloffs et al. (1995)
Matilija Cr	Kern County 21 Jul 1952	7.5	31	Muir-Wood and King (1993)

(continued)

Stream/spring	Earthquake	Magnitude	Distance (km)	References
Santa Cruz Cr	Kern County 21 Jul 1952	7.5	38	Muir-Wood and King (1993)
Grapevine Cr	Kern County 21 Jul 1952	7.5	46	Muir-Wood and King (1993)
Tunis Cr	Kern County 21 Jul 1952	7.5	48	Muir-Wood and King (1993)
Piru Cr	Kern County 21 Jul 1952	7.5	52	Muir-Wood and King (1993)
Canatsey-O'Bannon Sp	Kern County 21 Jul 1952	7.5	57	Muir-Wood and King (1993)
Sespe Cr	Kern County 21 Jul 1952	7.5	63	Manga et al. (2003)
El Paso Cr	Kern County 21 Jul 1952	7.5	71	Muir-Wood and King (1993)
Caliente Cr	Kern County 21 Jul 1952	7.5	83	Muir-Wood and King (1993)
Walker Basin Cr	Kern County 21 Jul 1952	7.5	99	Muir-Wood and King (1993)
	Kern County 21 Jul 1952	7.5	117, 131	Briggs and Troxell (1955) as reported in Montgomery et al. (2003)
Various	San Francisco 18 Apr 1906	7.9	37, 41, 48, 66, 95, 115, 118, 136, 148, 161, 166, 170, 178, 196, 205, 213, 246, 275, 340, 407	Lawson (1908) as compiled in Montgomery et al. (2003)
Ship Creek	Alaska 27 Mar 1964	9.2	92	Waller (1966)
South Fork, Campbell Crreek	Alaska 27 Mar 1964	9.2	121	Waller (1966)
	Alaska 27 Mar 1964	9.2	132, 320	Waller (1966) as reported in Montgomery et al. (2003)

Note a) Distance may not always be defined in a consistent manner. Following Manga (2001), distance is the distance between the epicenter and an estimate of the center of the gauged basin. Exceptions are all data for which the reference is given with the expression "as reported by"; the data for the Kobe earthquake and in Montgomery at al. (2003) are distances to the stream gauge or spring.

References

Briggs, R.O., 1991, Effects of Loma Prieta earthquake on surface waters in Waddell Valley, *Water Resour. Bull., 27*, 991–999.

Briggs, R.C., and H.C. Troxell, 1955 Effects of the Arvin-Tehachapi earthquake on spring and stream flow. In: G.B. Oakeshott (ed.), *Earthquakes in Kern County, Calif. Div. Mines and Geol. Bull., 171*, 81–97.

King, C.-Y., D. Basler, T.S. Presser, C.W. Evans, L.D. White, and A.D. Minissale, 1994, In search of earthquake-related hydrologic and chemical changes along the Hayward fault, *Appl. Geochem., 9*, 83–91.

Lawson, A.C., 1908, The California earthquake of April 18, 1906, Report of the State earthquake investigation commission, Vol. 1, Washington DC: Carnegie Institution of Washington.

Manga, M., 2001, Constraints on the origin of coseismic and postseismic streamflow changes inferred from recession-flow analysis, *Geophys. Res. Lett., 28*, 2133–2136.

Manga, M., E.E. Brodsky, and M. Boone, 2003, Response of streamflow to multiple earthquakes and implications for the origin of postseismic discharge changes, *Geophys. Res. Lett., 30*, doi:10.1029/2002GL016618.

McPherson, R.C. and L.A. Dengler, 1992, The Honeydew earthquake, August 12, 1991, *Calif. Geology, 45*, 31–39.

Montgomery, D.R., H.M. Greenberg, and D.T.Smith, 2003, Streamflow response to the Nisqually earthquake, *Earth Planet. Sci. Lett., 209*, 19–28.

Muir-Wood, R., and G.C.P. King, 1993, Hydrological signatures of earthquakes strain, *J. Geophys. Res., 98*, 22035–22068.

Quilty, E.G., C.D. Farrar, D.L. Galloway, S.N. Hamlin, R.J. Laczniak, E.A. Roeloffs, M.L. Sorey, and D.E. Woodcock, 1995, Hydrologic effects associated with the January 14, 1994 Northridge, California, earthquake, USGS Open File Report 95-813.

Roeloffs, E., W.R. Danskin, C.D. Farrarr, D.L. Galloway, S.N. Hamlin, E.G. Quilty, H.M. Quinn, D.H. Schaefer, M.L. Sorey, and D.E. Woodcock , 1995, Hydrologic effects of the June 28, 1992 Landers earthquake, U.S. Geol. Surv. Open File Report 95-42.

Rojstaczer, S., S. Wolf, and R. Michel, 1995, Permeability enhancement in the shallow crust as a cause of earthquake-induced hydrological changes, *Nature, 373*, 237–239.

Rowland, J.C., M. Manga, T.P. Rose, 2008, The influence of poorly interconnected fault zone flow paths on spring geochemistry, *Geofluids, 8*, 93–101.

Sato, T., R. Sakai, K. Furuya, and T. Kodama, 2000, Coseismic spring flow changes associated with the 1995 Kob earthquake, *Geophys. Res. Lett., 27*, 1219–1222.

Waller, R., 1966, Effects of the March 1964 Alaska earthquake on the hydrology of the Anchorage area, US Geol. Surv., Prof. Paper 544B.

Wang, C.-Y., M. Manga, D. Dreger, and A. Wong (2004) Streamflow increases due to the rupturing of hydrothermal reservoir – evidence from the 2003 San Simeon earthquake (2003), California, Earthquake, *Geophys. Res. Lett., vol. 31*, L10502, doi:10.1029/2004GL020124.

E.2 Groundwater Level Responses

(a) A global data set, compiled by Wang and Chia (2008)

Earthquake	M	Well	Epicenter distance (km)	Δh (m)	References
1964 Alaska	9.2	Florida	5000	0.2	Cooper et al. (1965)
1964 Alaska	9.2	Hbt, Belgm	7215	0.01	Vorhis (1968)
1989 LomaPrieta	6.9	BV	157	0.85	Roeloffs (1998)
1992 Landers	7.3	BV	433	0.34	Roeloffs (1998)
1990 Chittenden	5.4	BV	133	0.05	Roeloffs (1998)
1992 Parkfield	4.7	BV	233	0.14	Roeloffs (1998)
1993 Parkfield	4.8	BV	269	0.36	Roeloffs (1998)
1994 Northridge	6.7	BV	281	0.19	Roeloffs (1998)
1994 Mendocino	7	BV	732	0.09	Roeloffs (1998)
1994 Parkfield	5	BV	24	0.33	Roeloffs (1998)
1989 LomaPrieta	6.9	LKT	272	–0.146	Roeloffs et al. (2003)
1992 Landers	7.3	LKT	451	–0.396	Roeloffs et al. (2003)
1999 HectorMine	7.1	LKT	421	–0.17	Roeloffs et al. (2003)
1992 Landers	7.3	CW-3	451	0.35	Roeloffs et al. (2003)
1999 HectorMine	7.1	CW-3	421	–0.12	Roeloffs et al. (2003)
1992 Landers	7.3	CH10-B	451	–0.07	Roeloffs et al. (2003)
1994 Petrolia	7	NVIP-3	300	–0.16	Brodsky et al. (2003)
1999 Oaxaca	7.4	NVIP-3	3850	–0.11	Brodsky et al. (2003)
2003 Tokachi-oki	8	Japan	1200	0.02	Sato et2004 lower
2003 Tokachi-oki	8	Japan	1200	–0.02	Sato et2004 upper
2004 Sumatra	9.3	Alaska	10800	0.001	Sil-Freymueller (2006)
2004 Sumatra	9.3	Japan	5000	0.03	Kitagawa et al. (2006)
2004 Sumatra	9.3	Japan	5000	0.03	Kitagawa et al. (2006)
2004 Sumatra	9.3	Japan	5000	0.03	Kitagawa et al. (2006)
2004 Sumatra	9.3	Japan	5000	0.03	Kitagawa et al. (2006)
2004 Sumatra	9.3	Japan	5000	0.03	Kitagawa et al. (2006)
2004 Sumatra	9.3	Japan	5000	0.03	Kitagawa et al. (2006)
2004 Sumatra	9.3	Japan	5000	0.03	Kitagawa et al. (2006)
2004 Sumatra	9.3	Japan	5000	0.03	Kitagawa et al. (2006)

(continued)

Earthquake	M	Well	Epicenter distance (km)	Δh (m)	References
2004 Sumatra	9.3	Japan	5000	–0.03	Kitagawa et al. (2006)
2004 Sumatra	9.3	Japan	5000	–0.03	Kitagawa et al. (2006)
2004 Sumatra	9.3	Japan	5000	–0.03	Kitagawa et al. (2006)
1981/8/15	4.8	Haibara, JP	41.9	–0.064	Matsumoto et al. (2003)
1982/7/23	7	Haibara, JP	375	–0.035	Matsumoto et al. (2003)
1982/12/28	6.4	Haibara, JP	155.9	–0.036	Matsumoto et al. (2003)
1983/3/16	5.7	Haibara, JP	65.9	–0.044	Matsumoto et al. (2003)
1983/5/26	7.7	Haibara, JP	621.9	–0.018	Matsumoto et al. (2003)
1983/8/8	6	Haibara, JP	113.1	–0.026	Matsumoto et al. (2003)
1983/11/24	5	Haibara, JP	57	–0.019	Matsumoto et al. (2003)
1984//9/14	6.8	Haibara, JP	127.9	–0.14	Matsumoto et al. (2003)
1986/6/24	6.5	Haibara, JP	241.7	–0.013	Matsumoto et al. (2003)
1986/11/22	6	Haibara, JP	126.2	–0.031	Matsumoto et al. (2003)
1986/12/17	6.7	Haibara, JP	226.7	–0.035	Matsumoto et al. (2003)
1989/10/14	5.7	Haibara, JP	122.1	–0.017	Matsumoto et al. (2003)
1990/02/20	6.5	Haibara, JP	95.9	–0.081	Matsumoto et al. (2003)
1990/9/24	6.6	Haibara, JP	199.7	–0.013	Matsumoto et al. (2003)
1991/4/25	4.9	Haibara, JP	43.9	–0.011	Matsumoto et al. (2003)
1991/9/3	6.3	Haibara, JP	139.9	–0.03	Matsumoto et al. (2003)
1993/7/12	7.8	Haibara, JP	888.8	–0.01	Matsumoto et al. (2003)
1994/10/4	8.1	Haibara, JP	1253.7	–0.074	Matsumoto et al. (2003)
1994/12/18	7.5	Haibara, JP	794.3	–0.013	Matsumoto et al. (2003)
1995/1/17	7.2	Haibara, JP	289.8	–0.039	Matsumoto et al. (2003)
1996/2/1	3.6	Haibara, JP	13.2	–0.008	Matsumoto et al. (2003)
1996/3/6	5.8	Haibara, JP	104.4	–0.022	Matsumoto et al. (2003)

(continued)

Earthquake	M	Well	Epicenter distance (km)	Δh (m)	References
1996/5/27	4.2	Haibara, JP	33.7	–0.031	Matsumoto et al. (2003)
1996/10/5	4.3	Haibara, JP	36.2	–0.011	Matsumoto et al. (2003)
1997/3/16	5.8	Haibara, JP	73.3	–0.088	Matsumoto et al. (2003)
1997/10/11	4.9	Haibara, JP	54.4	–0.062	Matsumoto et al. (2003)
1988/12/09	6.8	Haibara, JP	110	–0.75	Matsumoto et al. (2003)

(b) Responses to the 1999 Chi-Chi earthquake, compiled by Chia et al. (2008)

*The epicentral distance of the wells in the following two tables may be calculated as the square root of (dX*dX + dY*dY), where dX is the difference between the TM_X of the station and the TM_X of the epicenter, dY is the distance between the TM_Y of the well and the TM_Y of the epicenter, and the TM-coordinates are referenced to a local geo-reference system in Taiwan. The epicenter of the Chi-Chi earthquake in this coordinate system is (231665 m, 2638432 m). The number in the parenthesis following the well name identifies the aquifer beneath the surface (see Fig. 5.3).

Well	TM_X (m)	TM_Y (m)	Coseismic change (cm)
Anhe (2)	178980	2601660	9
Anhe (3)	178980	2601660	8
Anhe (4)	178980	2601660	27
Annan (1)	172567	2622644	82
Annan (2)	172567	2622644	144
Anping (2)	163950	2544815	4
Beigang (1)	177919	2608767	63
Beigang (2)	177919	2608767	97
Beimen (1)	158374	2576594	13
Beimen (2)	158374	2576580	33
Bozi (1)	162598	2614898	15
Bozi (2)	162598	2614898	56
Caicuo (1)	169500	2612504	12
Caicuo (2)	169500	2612504	52
Chaoliao (2)	190004	2496500	34
Chaoming (1)	188414	2493802	4
Chifeng (4)	197178	2480938	22
Chongjeng (2)	185015	2503504	–4
Chongxi (3)	184263	2576361	31
Chongxi (4)	184263	2576361	29
Chukou (1)	212782	2629816	–405
Chukou (2)	212782	2629816	–275
Dagou (1)	168595	2607406	14
Dagou (2)	168595	2607406	50
Dahu (2)	200368	2497500	15
Dahu (3)	200368	2497500	–59
Dahu (4)	200368	2497500	2
Datan (1)	197221	2484469	3
Datan (2)	197221	2484469	13
Dawen (3)	162586	2561292	3
Dongfang (1)	199950	2662265	338
Dongfang (2)	199950	2662265	373
Donggan (2)	193156	2485971	18
Donggan (3)	193156	2485971	16
Donggan (4)	193156	2485971	13
Dongguang (2)	174905	2616870	99

(continued)

Well	TM_X (m)	TM_Y (m)	Coseismic change (cm)
Dongguang (3)	174905	2616870	134
Dongguang (4)	174905	2616870	248
Dongguang (5)	174905	2616870	176
Donghe (3)	205251	2620504	–19
Dongrong (1)	191545	2606340	28
Dongrong (2)	191545	2606340	–9
Dongrong (4)	191545	2606340	14
Dongshi (1)	162790	2595720	22
Dongshi (2)	162790	2595720	48
Dongshi (3)	162790	2595720	35
Dongshi (4)	162790	2595720	22
Fangcao (1)	185350	2624184	18
Fangcao (2)	185350	2624184	580
Fangyuan (1)	179990	2646954	141
Fangyuan (2)	179990	2646954	54
Fangyuan (3)	179990	2646954	141
Fengrong (1)	178954	2632222	18
Fengrong (2)	178954	2632222	182
Fengrong (3)	178954	2632222	106
Gangdong (2)	195798	2490241	5
Gangdong (3)	195798	2490241	5
Gangdong (4)	195798	2490241	7
Ganghou (1)	187220	2632820	321
Ganghou (3)	187220	2632820	367
Ganghou (4)	187220	2632820	271
Gangwei (1)	167942	2567431	–7
Gangwei (2)	167942	2567431	9
Gangwei (3)	167942	2567431	29
Guantian (2)	181773	2565588	4
Guantian (3)	181773	2565588	4
Gukeng (1)	204980	2615932	–74
Guosheng (1)	205365	2665558	38
Guosheng (2)	205365	2665558	521
Guosheng (3)	205365	2665558	105
Haifeng (1)	170273	2629402	17
Haifeng (2)	170273	2629402	68
Haifeng-a (1)	198476	2511130	–3
Haifeng-a (3)	198476	2511130	–4
Haiyuan (1)	165467	2624551	23
Haiyuan (2)	165467	2624551	42
Haiyuan (3)	165467	2624551	47
Haiyuan (4)	165467	2624551	24
Hanbao (1)	183282	2656147	256
Hanbao (2)	183282	2656147	323
Hanbao (3)	183282	2656147	212
Hanbao (4)	183282	2656147	125
Haoxiu (1)	194052	2656100	351
Haoxiu (2)	194052	2656100	528

(continued)

Well	TM_X (m)	TM_Y (m)	Coseismic change (cm)
Haoxiu (3)	194052	2656100	372
Haoxiu (4)	194052	2656100	324
Hefeng (1)	169999	2626544	39
Hefeng (2)	169999	2626544	74
Hexing (1)	194000	2643600	11
Hexing (2)	194000	2643600	443
Hexing (3)	194000	2643600	235
Honglun (1)	182680	2620675	37
Honglun (2)	182680	2620675	455
Houan (2)	171197	2632096	48
Huatan (1)	202725	2658257	125
Huatan (2)	202725	2658257	131
Huatan (3)	202725	2658257	398
Huatan (4)	202725	2658257	–26
Huwei (2)	191285	2623689	742
Huxi (1)	199331	2624542	83
Huxi (2)	199331	2624542	257
Huxi (3)	199331	2624542	252
Huxi (4)	199331	2624542	268
Jende (2)	172760	2540572	–4
Jian (2)	171010	2561511	24
Jiaxing (1)	194034	2616371	167
Jinhu (1)	164163	2577533	8
Jinhu (2)	164163	2577533	5
Jinxue (3)	167156	2558381	19
Jinxue (4)	167156	2558381	18
Jiulong (1)	191168	2627781	88
Jiulong (2)	191168	2627781	411
Jiulong (3)	191168	2627781	504
Jiuru (1)	196776	2515370	–6
Jiuru (2)	196776	2515370	3
Jiuzhuang (2)	188020	2614870	41
Jiuzhuang (3)	188020	2614870	129
Jiuzhuang (4)	188020	2614870	241
Jiuzhuang (5)	188020	2614870	197
Kanding (2)	198929	2490845	7
Kanding (3)	198929	2490845	4
Kanjiao (1)	202021	2612378	50
Kanjiao (2)	202021	2612378	29
Kinghu (1)	162471	2608474	18
Kinghu (2)	162471	2608474	37
Ligang (1)	197514	2520447	6
Ligang (2)	197514	2520447	5
Linyuan (2)	186809	2490130	4
Linyuan (3)	186809	2490130	5
Littlexin (1)	178501	2558459	–14
Liuing (2)	178870	2575008	56
Liujia (1)	182708	2569605	–3

(continued)

Well	TM_X (m)	TM_Y (m)	Coseismic change (cm)
Liujia (2)	182708	2569605	5
Liujia (3)	182708	2569605	–44
liyu (1)	213977	2623841	–462
liyu (2)	213977	2623841	–594
Lunzi (1)	183331	2611929	30
Lunzi (2)	183331	2611929	214
Luojin (1)	191220	2661365	62
Luojin (2)	191220	2661365	522
Luojin (3)	191220	2661365	347
Majia (1)	208976	2511539	–3
Mingde (1)	167485	2617020	10
Mingde (2)	167485	2617020	74
Mingde (3)	167485	2617020	102
Mingde (4)	167485	2617020	75
Nanxing (2)	169053	2553702	10
Nanxing (3)	169053	2553702	26
Nanxing (4)	169053	2553702	33
Neipu (1)	204588	2501296	–10
Neipu (2)	204588	2501296	–4
Newhua (1)	178148	2549104	–4
Newhua (2)	178148	2549104	–18
Newshi (1)	177117	2553431	–7
Newshi (2)	177117	2553431	10
Newshi (3)	177117	2553431	16
Newyuan (2)	192774	2491800	12
Pingding (1)	212545	2628234	–87
Pingxi (1)	169486	2584223	–6
Pingxi (2)	169486	2584223	45
Pingxi (4)	169486	2584223	23
Qiongpu (1)	168219	2602115	20
Qiongpu (2)	168219	2602115	20
Quanxing (1)	199630	2674365	19
Quanxing (2)	199630	2674365	466
Quanxing (3)	199630	2674365	386
Quanxing (4)	199630	2674365	270
Sangu (1)	158802	2556075	5
Sangu (2)	158802	2556075	4
Sanhe (1)	196915	2611594	–37
Sanhe (2)	196915	2611594	29
Shaying (2)	173545	2570238	30
Shaying (3)	173545	2570238	51
Shaying (4)	173545	2570238	51
Shifen (1)	154146	2556442	6
Shifen (2)	154146	2556442	9
Shifen (3)	154146	2556442	14
Shihua (2)	187480	2487970	9
Shipu (2)	191978	2513893	–4
Shipu (3)	191978	2513893	–7

(continued)

Well	TM_X (m)	TM_Y (m)	Coseismic change (cm)
Shuilin (1)	172216	2608151	28
Singbei (3)	202895	2485957	–4
Singbei (4)	202895	2485957	4
Tanqian (1)	182705	2637176	81
Tanqian (2)	182705	2637176	401
Tianwei (1)	201045	2643280	203
Tianwei (2)	201045	2643280	395
Tianyang (2)	178717	2624989	72
Tianyang (3)	178717	2624989	95
Tianzhong (2)	207088	2639188	–247
Tzeshan (3)	209521	2499542	–3
Wanlong (2)	206802	2490112	–3
Wenchang (2)	190120	2656250	276
Wenchang (3)	190120	2656250	409
Wenchang (4)	190120	2656250	338
Wencuo (2)	199400	2617408	32
Xiangtian (2)	185725	2641405	159
Xianxi (1)	195062	2669966	23
Xianxi (2)	195062	2669966	420
Xianxi (3)	195062	2669966	428
Xianxi (4)	195062	2669966	275
Xigang	176799	2639978	247
Xigang (1)	176799	2639978	60
Xigang (2)	176799	2639978	269
Xigang (3)	176799	2639978	251
Xigang (4)	176799	2639978	119
Xihu (1)	196133	2649778	417
Xihu (2)	196133	2649778	476
Xihu (3)	196133	2649778	515
Xindong (1)	179177	2582776	–9
Xindong (2)	179177	2582776	90
Xindong (3)	179177	2582776	34
Xindong (4)	179177	2582776	56
Xinghua (1)	176695	2628849	57
Xinghua (2)	176695	2628849	113
Xinghua (3)	176695	2628849	174
Xinguang (1)	215584	2638740	–911
Xinmin (1)	218500	2634400	19
Xishi (3)	200397	2502704	–7
Xishi (4)	200397	2502704	–30
Xizhou (1)	198371	2639267	16
Xizhou (2)	198371	2639267	475
Xizhou (3)	198371	2639267	70
Yiwu (2)	166296	2604660	21
Yiwu (3)	166296	2604660	51
Yiwu (4)	166296	2604660	61
Yongfang (3)	186942	2500950	5
Yongkang (2)	172843	2547566	4

(continued)

Well	TM_X (m)	TM_Y (m)	Coseismic change (cm)
Yuanlin (1)	205885	2649930	655
Yuanlin (2)	205885	2649930	646
Yuanlin (3)	205885	2649930	412
Yuanlin (4)	205885	2649930	120
Yuanzhang (1)	178788	2616961	176
Zhaojia (1)	187624	2648441	464
Zhaojia (2)	187624	2648441	540
Zhaojia (3)	187624	2648441	439
Zhongzhou (1)	195993	2528009	23
Zhongzhou (2)	195993	2528009	4
Zhushan (1)	217282	2629020	–712
Zhushan (2)	217282	2629020	–1109
Zhutang (2)	190948	2639823	386
Zhuwei (1)	171588	2595629	31
Zhuwei (2)	171588	2595629	65
Zhuwei (3)	171588	2595629	45
Zongye (1)	174375	2565395	–26
Zongye (2)	174375	2565395	9

(c) Responses to the 2006 Hengchun earthquake, compiled by Chia et al. (2008)

* The epicenter of the Hengchun earthquake in the Taiwan coordinate system is (190100 m, 2430357m).

Well	TM_X (m)	TM_Y (m)	Coseismic change (cm)
Anching (1)	166368	2549095	1
Anching (2)	166368	2549095	–1
Anching (3)	166368	2549095	–5
Anching (4)	166368	2549095	–14
Caicuo (2)	169500	2612504	–4
Chaoliao (3)	190004	2496500	–4
Chaoming (1)	188414	2493802	17
Chaoming (2)	188414	2493802	–9
Chaoming (3)	188414	2493802	–10
Chaoming (4)	188414	2493802	–10
Chifeng (2)	197199	2480945	16
Chifeng (3)	197199	2480945	22
Chifeng (4)	197199	2480945	23
Chingxi (3)	194408	2506330	–3
Chongjeng (1)	185015	2503504	–8
Chongjeng (2)	185015	2503504	–5
Chukou (2)	212782	2629816	–3
Dacer (1)	181950	2514800	6

(continued)

Well	TM_X (m)	TM_Y (m)	Coseismic change (cm)
Dahu (2)	200368	2497500	14
Dalun (2)	184712	2595490	–5
Dashu (3)	191135	2510159	–5
Dashu (4)	191135	2510159	–3
Datan (2)	197221	2484469	–4
Dawen (1)	162586	2561292	2
Dawen (2)	162586	2561292	–3
Dawen (3)	162586	2561292	–2
Daxiang (2)	210206	2481360	7
Donggan (2)	193156	2485971	29
Donggan (3)	193156	2485971	50
Donghe (2)	205251	2620504	–10
Fangcao (2)	185350	2624184	2
Fanhua (2)	205264	2511656	1
Fengming (1)	182982	2490949	4
Gangdong (3)	195798	2490241	–4
Gangdong (4)	195798	2490241	–7
Ganghe (3)	181125	2496875	–7
Gangqian (1)	168959	2596050	4
Gangqian (2)	168959	2596050	–3
Gangshan (1)	174100	2524520	11
Gangwei (3)	167942	2567431	–6
Guanfu (2)	209883	2518620	2
Guantian (3)	181773	2565588	–3
Haifeng (2)	198476	2511130	8
Honglun (2)	182680	2620675	–4
Jende (2)	172760	2540572	–3
Jian (2)	171010	2561511	–7
Jianxing (2)	204395	2507086	7
Jiuqu (1)	190015	2506507	–29
Jiuqu (2)	190015	2506507	–23
Linyuan (3)	186809	2490130	–2
Littlexin (2)	178501	2558459	–2
Liuing (2)	178870	2575008	–6
Liujia (1)	182708	2569605	–2
Liujia (2)	182708	2569605	–8
Majia (3)	208976	2511539	2
Naba (2)	182416	2552332	–4
Nanke (2)	175430	2556230	–9
Nanke (3)	175430	2556230	–3
Neipu (2)	204588	2501296	10
Newshi (1)	177117	2553431	–17
Newyuan (2)	192774	2491800	–4
Pengtsuo (2)	201104	2515658	2
Pingding (1)	212545	2628234	2
Pingxi (2)	169486	2584223	–5
Pingxi (3)	169486	2584223	2
Qianjin (1)	193652	2505845	–3

(continued)

Well	TM_X (m)	TM_Y (m)	Coseismic change (cm)
Shangtan (1)	208707	2486714	14
Shangtan (2)	208707	2486714	–2
Shanwa (1)	176200	2559721	3
Shanwa (2)	176200	2559721	–2
Shaying (2)	173545	2570238	–8
Shaying (4)	173545	2570238	–7
Sheliao (2)	219363	2440038	54
Sheliao (3)	219363	2440038	24
Shifen (2)	154146	2556442	–3
Shipu (3)	191978	2513893	–3
Singbei (2)	202895	2485957	29
Singbei (4)	202895	2485957	9
Taimount (2)	209266	2521392	2
Tainan (1)	160526	2548099	7
Tainan (4)	160526	2548099	2
Texing (3)	207000	2474613	28
Tsaojoe (2)	202232	2492984	11
Tzeshan (1)	209521	2499542	6
Tzeshan (3)	209521	2499542	–18
Wandan (2)	194540	2501970	9
Wandan (3)	194540	2501970	–3
Wanlong (2)	206802	2490112	10
Wanrun (2)	207502	2497136	10
Wujia (1)	181028	2540128	–14
Wuwoods (2)	176549	2518073	–5
Wuwoods (3)	176549	2518073	–3
Xiangtian (1)	185725	2641405	–7
Xishi (2)	200397	2502704	25
Xishi (3)	200397	2502704	21
Xishi (4)	200397	2502704	12
Yancheng (1)	175440	2502825	9
Yancheng (3)	175440	2502825	–3
Yanpu (2)	205622	2517240	3
Yongfang (2)	186942	2500950	–4
Yongfang (3)	186942	2500950	–12
Yonghua (1)	169575	2524125	4
Zhushan (2)	217282	2629020	2
Zongye (3)	174375	2565395	–6

References

Brodsky, E.E., E. Roeloffs, D. Woodcock, I. Gall, and M. Manga, 2003, A mechanism for sustained groundwater pressure changes induced by distant earthquakes, *J. Geophys. Res., 108*, 2390, doi:10.1029/2002JB002321.

Chia, Y., J.J. Chiu, Y.H. Jiang, T.P. Lee, Y.M. Wu, and M.J. Horng, 2008, Implications of coseismic groundwater level changes observed at multiple-well monitoring stations, *Geophys. J. Int., 172,* 293–301.

Cooper, H.H., J.D. Bredhoeft, I.S. Papdopulos, and R.R. Bennnett, 1965, The response of aquifer-well systems to seismic waves, *J. Geophys. Res., 70*, 3915–3926.

Kitagawa, Y., N. Koizumi, M. Takahashi, N. Matsumoyo, and T. Sato, 2006, Changes in water levels or pressures associated with the 2004 earthquake off the west coast of northern Sumatra (M9.0), *Earth Planets Space, 58*, 173–179.

Matsumoto, N., G. Kitagawa, and E.A. Roeloffs, 2003, Hydrological response to earthquakes in the Haibara well, central Japan – I. Water level changes revealed using state space decomposition of atmospheric pressure, rainfall and tidal responses, *Geophys. J. Int., 155*, 885–898.

Roeloffs, E.A., 1998, Persistent water level changes in a well near Parkfield, California, due to localand distant earthquakes, *J. Geophys. Res., 103*, 869–889.

Roeloffs, E.A., M. Sneed, D.L. Galloway, M.L. Sorey, C.D. Farrar, J.F. Howle, J. Hughes, 2003, Water-level changes induced by local and distant earthquakes at Long Valley caldera, California, *J. Volcan. Geotherm. Res., 127*, 269–303.

Sato, T., N. Matsumoto, Y. Kitagawa, N. Koizumi, M. Takahashi, Y. Kuwahara, H. Ito, A. Cho, T. Satoh, K. Ozawa, and S. Tasaka, 2004, Changes in water level associated with the 2003 Tokachi-oki earthquake, *Earth Planets Space, 56*, 395–400.

Sil, S., and J.T. Freymueller, 2006, Well water level changes in Fairbanks, Alaska, due to the great Sumatra-Andaman earthquake, *Earth Planets Space, 58*, 181–184.

Vorhis, R.C., 1968, Effects outside Alaska, in *The Great Alaska of 1964, Hydrology, Nat. Acad. Sc. Publ., 1603*, Washington DC, 140–189.

Wang, C., and Y. Chia, 2008, Mechanism of water level changes during earthquakes: Near field versus intermediate field, *Geophys. Res. Lett., 35*, L12402, doi:10.1029/2008GL034227.

E.3 Hot Spring Responses

Earthquake	M	Date	Epicenter distance (km)	Focal depth (km)
Ito-oki	2.3	1982/7/14	6	0
Ito-oki	4.5	1984/9/5	16	12
Ito-oki	2.8	1985/1/20	18	8
Ito-oki	3.6	1985/10/21	2	5
Ibaraki-ken-oki	7.0	1982/7/23	290	30
Sagami-bay	5.7	1982/8/12	46	30
Torishima	7.9	1984/3/6620	452	
Boso-hanto-oki	6.5	1986/6/24	152	80

Reference

Mogi, K., H. Mochizuki, and Y. Kurokawa, 1989, Temperature changes in an artesian spring at Usami in the Izu Peninsula (Japan) and their relation to earthquakes, *Tectonophysics, 159*, 95–108.

E.4 Liquefaction Occurrence During Earthquakes

(a) Distance to **liquefaction and focal depth, compiled by Ambraseys (1989)**

Country	Date	M	Epicenter distance (km)	Focal depth(km)
NZ	1848/10/15	7.1	126	40
NZ	1855/1/23	7.6	168	20
GR	1861/12/26	6.7	25	15
JP	1872/3/14	7.1	107	20
SP	1884/12/15	6.8	45	15
JP	1887/7/22	5.7	6	5
NZ	1888/8/31	6.9	50	25
JP	1889/7/28	6.3	24	10
JP	1891/10/27	7.8	110	20
JP	1894/6/20	7	47	15
JP	1894/10/22	6.7	35	10
JP	1895/1/18	7.2	67	15
SU	1895/7/8	7.5	150	30
JP	1896/8/31	7.2	98	10
JP	1897/1/16	5.3	4	5
AU	1897/5/10	6.5	41	14
IN	1897/6/12	8.3	325	25
JP	1898/8/10	6	10	5
NZ	1901/11/15	7.3	65	10
GR	1902/7/5	6.4	20	11
BU	1904/4/4	7.4	135	18
IN	1905/4/4	8.2	210	25
US	1906/4/18	7.9	310	15
JM	1907/1/14	6.3	30	10
PR	1909/4/23	6.6	36	12
JP	1909/8/14	6.9	48	10
TR	1912/8/9	7.3	120	16
JP	1914/3/14	6.7	17	5
IN	1918/7/8	7.6	116	15
JP	1922/12/7	6.7	25	10
JP	1923/9/1	7.9	140	25
JP	1925/5/23	6.7	27	8
JP	1927/3/7	7.1	110	13
JP	1927/10/27	5	2	5
NZ	1929/3/9	6.9	35	30
IN	1930/7/2	7.1	68	20
JP	1930/10/12	6.2	19	5
NZ	1931/2/2	7.7	140	20
JU	1931/3/8	6.7	40	5
JP	1931/9/21	6.6	60	10
GR	1932/9/26	6.9	52	6
US	1933/3/11	6.3	20	10
JP	1933/9/21	5.8	5	5
IN	1934/1/15	8.1	305	30

(continued)

Country	Date	M	Epicenter distance (km)	Focal depth(km)
US	1934/3/12	6.6	30	15
PK	1935/5/30	7.5	150	20
JP	1935/7/11	5.9	9	10
US	1935/10/19	6.3	24	15
JP	1936/2/21	6.4	21	10
JP	1936/11/2	7.4	131	50
TR	1938/4/19	6.6	37	15
JP	1939/5/1	6.7	30	15
US	1940/5/19	7	70	7
RO	1940/11/10	7.5	350	135
JP	1941/7/15	6	8	5
NZ	1942/6/24	6.9	63	20
JP	1943/3/4	6	12	10
JP	1943/9/10	7	24	13
AR	1944/1/15	7.2	50	15
JP	1944/12/7	8	225	20
JP	1945/1/12	6.6	42	11
CN	1946/6/23	7.2	100	25
SU	1946/11/2	7.5	103	30
SU	1946/11/4	6.9	48	25
JP	1946/12/10	8.1	261	20
JP	1948/6/28	7.3	31	14
IN	1950/8/15	8.3	290	30
JP	1952/3/4	8.1	153	25
JP	1952/3/7	6.5	34	15
IR	1953/2/12	6.5	35	10
FJ	1953/9/14	6.4	96	60
US	1954/7/6	6.3	28	10
JP	1955/10/19	5.7	5	5
US	1957/3/22	5.3	5	7
US	1958/4/7	7	92	10
US	1958/7/10	7.9	245	15
MX	1959/8/26	6.4	38	20
MR	1960/2/29	5.7	6	5
CH	1960/5/22	9.5	370	25
JP	1961/2/1	5.2	2	5
JP	1961/2/26	7.3	95	56
JP	1962/4/30	6.4	20	5
IR	1962/9/1	7.1	20	15
JU	1963/7/26	6.1	0	5
US	1964/3/28	9.2	480	20
JP	1964/6/16	7.5	95	12
CH	1965/3/28	7.2	125	68
SS	1965/5/3	5.8	7	23
UG	1966/3/20	6.9	46	15
TR	1966/8/19	6.8	53	15
GR	1966/10/29	5.8	7	6
TR	1967/7/22	7	65	9

(continued)

Country	Date	M	Epicenter distance (km)	Focal depth(km)
VZ	1967/7/30	6.5	51	16
JP	1968/2/21	5.7	6	3
US	1968/4/9	6.5	56	11
JP	1968/5/16	8.2	286	20
NZ	1968/5/23	7.1	34	12
IR	1968/8/31	7.3	22	15
MX	1968/9/25	5.7	54	138
IN	1969/4/13	5.7	9	5
SA	1969/9/29	6.3	23	9
US	1969/10/1	4.8	0	8
IN	1970/3/23	5.7	8	3
PR	1970/5/31	7.9	107	50
NG	1971/1/10	7.8	125	41
US	1971/2/9	6.6	15	9
IR	1972/4/10	6.8	36	12
NC	1972/12/23	6.2	0	5
MX	1973/1/30	7.6	187	48
US	1973/2/21	5.6	7	8
CR	1973/4/13	5.7	10	10
MX	1973/8/28	7.1	210	75
GR	1973/11/4	5.7	0	13
LI	1974/10/8	7.1	34	41
CA	1975/2/4	7	95	16
SI	1975/7/20	7.3	44	54
GR	1975/12/31	5.8	7	4
GU	1976/2/4	7.4	220	5
IT	1976/5/6	6.4	15	12
CA	1976/7/27	7.6	180	10
RO	1977/3/4	7.5	310	91
AR	1977/11/23	7.2	102	5
JP	1978/1/14	6.7	32	4
JP	1978/2/20	6.4	107	60
JP	1978/6/12	7.5	120	48
GR	1978/6/20	6.4	13	5
US	1978/8/13	5.7	0	9
JU	1979/4/15	6.9	60	10
US	1979/10/15	6.7	20	12
EC	1979/12/12	7.8	202	28
AL	1980/10/10	7.2	26	8
IT	1980/11/23	6.9	0	13
US	1981/4/26	6	12	6
JP	1983/5/26	7.6	130	24
US	1983/10/28	7	56	10
GN	1983/12/22	6.3	36	11
MX	1980/6/9	6.4	28	5

(b) Distance to liquefaction up to 1980, compiled by Galli (2000)

Country	Date	M	Epicenter distance (km)
IT	1117/1/3	6.4	89
IT	1505/1/3	5	3
IT	1542/6/13	6.2	6
IT	1542/12/10	6.4	36
IT	1545/6/9	5.2	13
IT	1561/8/19	6.4	26
IT	1570/11/17	5.5	7
IT	1570/11/17	5.5	1
IT	1570/11/17	5.5	21
IT	1570/11/17	5.5	5
IT	1570/11/17	5.5	3
IT	1570/11/17	5.5	.
IT	1570/11/17	5.5	3
IT	1570/11/17	5.5	0
IT	1570/11/17	5.5	3
IT	1624/3/18	5.5	4
IT	1627/7/30	7	21
IT	1627/7/30	7	15
IT	1627/7/30	7	14
IT	1627/7/30	7	.
IT	1627/7/30	7	15
IT	1627/7/30	7	40
IT	1638/3/27	7.3	17
IT	1638/3/27	7.3	15
IT	1638/3/27	7.3	.
IT	1688/6/5	7.3	19
IT	1688/6/5	7.3	28
IT	1693/1/11	7.3	7
IT	1693/1/11	7.3	21
IT	1693/1/11	7.3	71
IT	1693/1/11	7.3	32
IT	1693/1/11	7.3	23
IT	1693/1/11	7.3	3
IT	1693/1/11	7.3	28
IT	1703/2/2	6.2	7
IT	1703/2/2	6.2	9
IT	1753/3/9	5.5	.
IT	1781/4/4	6.2	9
IT	1781/4/4	6.2	.
IT	1781/4/4	6.2	5
IT	1781/4/4	6.2	2
IT	1783/2/5	7.3	34
IT	1783/2/5	7.3	15
IT	1783/2/5	7.3	25
IT	1783/2/5	7.3	25
IT	1783/2/5	7.3	2

(continued)

Country	Date	M	Epicenter distance (km)
IT	1783/2/5	7.3	4
IT	1783/2/5	7.3	87
IT	1783/2/5	7.3	17
IT	1783/2/5	7.3	16
IT	1783/2/5	7.3	11
IT	1783/2/5	7.3	34
IT	1783/2/5	7.3	5
IT	1783/2/5	7.3	36
IT	1783/2/5	7.3	9
IT	1783/2/5	7.3	16
IT	1783/2/5	7.3	8
IT	1783/2/5	7.3	40
IT	1783/2/5	7.3	7
IT	1783/2/5	7.3	32
IT	1783/2/5	7.3	31
IT	1783/2/5	7.3	39
IT	1783/2/5	7.3	14
IT	1783/2/5	7.3	32
IT	1783/2/5	7.3	25
IT	1783/2/5	7.3	23
IT	1783/2/5	7.3	4
IT	1783/2/5	7.3	66
IT	1783/2/5	7.3	18
IT	1783/2/5	7.3	19
IT	1783/2/5	7.3	39
IT	1783/2/5	7.3	26
IT	1783/2/5	7.3	1
IT	1783/2/5	7.3	2
IT	1783/2/5	7.3	1
IT	1783/2/5	7.3	9
IT	1783/2/5	7.3	15
IT	1783/2/5	7.3	15
IT	1783/2/5	7.3	6
IT	1783/2/5	7.3	35
IT	1783/2/5	7.3	38
IT	1783/2/5	7.3	22
IT	1783/2/5	7.3	22
IT	1783/2/5	7.3	19
IT	1783/2/5	7.3	76
IT	1783/2/5	7.3	111
IT	1783/2/5	7.3	6
IT	1783/2/5	7.3	8
IT	1783/2/5	7.3	6
IT	1783/2/5	7.3	7
IT	1783/2/5	8	
IT	1783/2/5	7.3	5
IT	1783/2/5	7.3	6
IT	1783/2/5	7.3	6

(continued)

Country	Date	M	Epicenter distance (km)
IT	1783/2/5	7.3	5
IT	1783/2/5	7.3	10
IT	1783/2/5	7.3	45
IT	1783/2/5	7.3	40
IT	1783/2/5	7.3	4
IT	1783/2/5	7.3	4
IT	1783/2/5	7.3	28
IT	1783/2/5	7.3	1
IT	1783/2/5	7.3	5
IT	1783/2/5	7.3	2
IT	1783/2/5	7.3	42
IT	1783/2/7	7	11
IT	1783/2/7	7	71
IT	1783/2/7	7	4
IT	1783/2/27	5	2
IT	1783/3/28	6.7	19
IT	1783/3/28	6.7	7
IT	1783/3/28	6.7	8
IT	1783/3/28	6.7	14
IT	1783/3/28	6.7	20
IT	1783/3/28	6.7	32
IT	1783/3/28	6.7	11
IT	1783/3/28	6.7	47
IT	1783/3/28	6.7	22
IT	1783/3/28	6.7	5
IT	1783/3/28	6.7	18
IT	1783/3/28	6.7	11
IT	1783/3/28	6.7	14
IT	1783/3/28	6.7	12
IT	1783/3/28	6.7	28
IT	1783/3/28	6.7	12
IT	1783/3/28	6.7	6
IT	1783/3/28	6.7	6
IT	1783/3/28	6.7	12
IT	1783/3/28	6.7	34
IT	1785/10/9	5.5	4
IT	1785/10/9	5.5	4
IT	1785/10/13	5.5	10
IT	1786/12/25	5.5	10
IT	1789/9/30	5.9	1
IT	1802/5/15	5.5	6
IT	1802/5/15	5.5	.
IT	1805/7/26	6.7	.
IT	1805/7/26	6.7	.
IT	1805/7/26	6.7	2
IT	1805/7/26	6.7	105
IT	1805/7/26	6.7	6
IT	1805/7/26	6.7	20

(continued)

Country	Date	M	Epicenter distance (km)
IT	1818/2/20	6.2	10
IT	1818/2/20	6.2	25
IT	1818/2/20	6.2	22
IT	1826/2/1	5.2	8
IT	1832/1/13	5.9	3
IT	1832/1/13	5.9	4
IT	1832/1/13	5.9	4
IT	1832/1/13	5.9	.
IT	1832/3/8	6.4	15
IT	1832/3/8	6.4	10
IT	1832/3/8	6.4	14
IT	1832/3/13	5.2	11
IT	1836/4/25	6.4	10
IT	1846/8/14	5.9	.
IT	1846/8/14	5.9	2
IT	1846/8/14	5.9	2
IT	1846/8/14	5.9	2
IT	1846/8/14	5.9	2
IT	1846/8/14	5.9	6
IT	1846/8/14	5.9	14
IT	1854/2/12	6.4	10
IT	1854/2/12	6.4	16
IT	1854/2/12	6.4	16
IT	1857/12/16	7	4
IT	1857/12/16	7	8
IT	1857/12/16	7	30
IT	1857/12/16	7	5
IT	1857/12/16	7	.
IT	1870/10/4	6.4	24
IT	1870/10/4	6.4	21
IT	1873/6/29	6.4	4
IT	1875/3/17	5.2	27
IT	1875/3/17	5.2	19
IT	1875/12/6	5.2	20
IT	1887/2/23	6.4	19
IT	1887/2/23	6.4	30
IT	1887/2/23	6.4	24
IT	1887/2/23	6.4	49
IT	1889/10/13	5	1
IT	1893/8/10	5.2	3
IT	1894/3/25	5	4
IT	1894/11/16	5.9	7
IT	1894/11/16	5.9	4
IT	1894/11/16	5.9	23
IT	1894/11/16	5.9	30
IT	1894/11/16	5.9	29
IT	1898/11/2	4.2	8
IT	1901/10/30	5.5	3

(continued)

Country	Date	M	Epicenter distance (km)
IT	1902/3/5	5	5
IT	1905/9/8	7.5	.
IT	1905/9/8	7.5	36
IT	1905/9/8	7.5	25
IT	1905/9/8	7.5	43
IT	1905/9/8	7.5	13
IT	1905/9/8	7.5	47
IT	1905/9/8	7.5	78
IT	1905/9/8	7.5	22
IT	1905/9/8	7.5	37
IT	1905/9/8	7.5	40
IT	1905/9/8	7.5	5
IT	1905/9/8	7.5	79
IT	1905/9/8	7.5	48
IT	1905/9/8	7.5	20
IT	1905/9/8	7.5	19
IT	1908/12/28	7.3	12
IT	1908/12/28	7.3	10
IT	1908/12/28	7.3	10
IT	1908/12/28	7.3	11
IT	1908/12/28	7.3	6
IT	1908/12/28	7.3	.
IT	1909/8/25	5.1	10
IT	1915/1/13	7	63
IT	1915/1/13	7	13
IT	1915/1/13	7	9
IT	1915/1/13	7	3
IT	1915/1/13	7	6
IT	1915/1/13	7	3
IT	1915/1/13	7	3
IT	1915/1/13	7	5
IT	1915/1/13	7	7
IT	1915/1/13	7	1
IT	1915/1/13	7	29
IT	1915/1/13	7	.
IT	1915/1/13	7	.
IT	1916/5/17	6	8
IT	1916/8/16	6	5
IT	1916/8/16	6	2
IT	1916/8/16	6	20
IT	1916/8/16	6	5
IT	1916/8/16	6	14
IT	1917/4/26	5.6	4
IT	1919/6/29	6.3	4
IT	1919/6/29	6.3	6
IT	1919/9/10	5.2	4
IT	1930/7/23	6.7	7
IT	1930/7/23	6.7	33

(continued)

Country	Date	M	Epicenter distance (km)
IT	1968/1/15	5.9	26
IT	1968/1/15	5.9	18
IT	1968/1/15	5.9	4
IT	1968/1/15	5.9	4
IT	1968/1/15	5.9	18
IT	1968/1/15	5.9	9
IT	1976/5/6	6.5	8
IT	1976/5/6	6.5	8
IT	1976/5/6	6.5	7
IT	1976/5/6	6.5	8
IT	1976/5/6	6.5	7
IT	1976/5/6	6.5	7
IT	1976/5/6	6.5	9
IT	1976/5/6	6.5	7
IT	1976/5/6	6.5	7
IT	1976/5/6	6.5	7
IT	1976/5/6	6.5	7
IT	1976/5/6	6.5	2
IT	1976/5/6	6.5	7
IT	1976/5/6	6.5	6
IT	1976/5/6	6.5	6
IT	1976/5/6	6.5	6
IT	1976/5/6	6.5	6
IT	1976/5/6	6.5	2
IT	1976/5/6	6.5	1
IT	1976/5/6	6.5	2
IT	1976/5/6	6.5	2
IT	1976/5/6	6.5	3
IT	1976/5/6	6.5	4
IT	1976/5/6	6.5	4
IT	1976/5/6	6.5	8
IT	1976/5/6	6.5	1
IT	1976/5/6	6.5	1
IT	1976/5/6	6.5	1
IT	1976/5/6	6.5	1
IT	1976/5/6	6.5	7
IT	1976/5/6	6.5	6
IT	1976/5/6	6.5	7
IT	1976/5/6	6.5	7
IT	1976/5/6	6.5	6
IT	1976/5/6	6.5	6
IT	1976/5/6	6.5	4
IT	1976/5/6	6.5	3
IT	1976/5/6	6.5	7
IT	1976/5/6	6.5	7
IT	1976/5/6	6.5	4
IT	1976/5/6	6.5	1
IT	1976/5/6	6.5	6

(continued)

Country	Date	M	Epicenter distance (km)
IT	1976/5/6	6.5	5
IT	1976/5/6	6.5	4
IT	1976/5/6	6.5	4
IT	1976/9/15	5.9	7
IT	1976/9/15	5.9	7
IT	1976/9/15	5.9	13
IT	1980/11/23	6.9	.
IT	1980/11/23	6.9	8
IT	1980/11/23	6.9	20
IT	1980/11/23	6.9	14
IT	1980/11/23	6.9	16
IT	1980/11/23	6.9	12
IT	1980/11/23	6.9	44
IT	1980/11/23	6.9	22
IT	1980/11/23	6.9	20
IT	1980/11/23	6.9	18
IT	1980/11/23	6.9	17
IT	1980/11/23	6.9	43
IT	1980/11/23	6.9	21
IT	1980/11/23	6.9	55
IT	1980/11/23	6.9	59
IT	1980/11/23	6.9	36
IT	1980/11/23	6.9	35
IT	1980/11/23	6.9	35
IT	1980/11/23	6.9	30
IT	1980/11/23	6.9	23
IT	1980/11/23	6.9	28
IT	1990/12/13	4.7	14

(c) Maximum distance to liquefaction from 1983 to 2006 (Wang et al., 2006)

Earthquake	Magnitude	Epicentral Distance (km)	Focal Depth (km)	Source
1983 Nihonkai-Chubu, Japan	7.7	160	15	1
1988 Udaipu Gahri, India	6.6	100	10	2
1989 Loma Prieta, California	7.1	93	18	Bardet and Kapuskar (1993)
1994 Northridge, California	6.7	50	19	3
1995 Manzanillo, Mexico	7.3	150	30	4

(continued)

Earthquake	Magnitude	Epicentral Distance (km)	Focal Depth (km)	Source
1995 Kobe, Japan	6.9	40	10	5
1999 Izmit, Turkey	7.8	61	17	Rothaus et al. (2004)
1999 Duzce, Turkey	7.5	56	10	6
1999 Chi-Chi, Taiwan	7.6	80	8	Yu et al. (2000)
2001 Gujarat, India	7.7	260	17	Rajendran et al. (2001)
2001 Nisqually, Washington	6.8	75	52	Pierepiekarz et al. (2001)
2002 Denali, Alaska	7.9	300	4.2	Kayen et al. (2002)
2003 Colima, Mexico	7.7	60	30	7

[1] http://www.ce.berkeley.edu/_hausler/sites/NKC001.pdf
[2] http://asc-india.org/gq/udaipur.htm
[3] www.lafire.com/famous_fires/940117_NorthridgeEarthquake/quake/02_EQE_geology.htm
[4] http://sun1.pue.upaep.mx/servs/carrs/GIIS/manzanillo.html
[5] http://www.jrias.or.jp/public/Hanshin_Earthquake/q1-2e.html
[6] http://geoinfo.usc.edu/turkey/
[7] http://geoinfo.usc.edu/gees/

References

Ambraseys, N.N., 1988, Engineering seismology, *Earthq. Eng. Struc. Dyn., 17*, 1–105.

Bardet, J. P., and M. Kapuskar, 1993, Liquefaction sand boil in San Francisco during 1989 Loma Prieta earthquake, *J. Geotech. Eng., 119*, 543–562.

Galli, P., 2000, New empirical relationships between magnitude and distance for liquefaction, *Tectonophys., 324*, 169–187.

Kayen, R., E. Thompson, D. Minasiah, B. Collins, E.R.S. Moss, N. Sitar, and G. Carver, 2002, Geotechnical reconnaissance of the November 3, 2002 M7.9 Denali fault earthquake, *Earthquake Spectra*, Special Issue on the M7.9 Denali Fault Earthquake of 3 November 2002, 1–27.

Pierepiekarz, M.R., D.B. Ballantyne, and R.O. Hamberger, 2001, Damage report from Seattle, *Civ. Eng. (Am. Soc. Civ. Eng.), 71*, 78–83.

Rajendran, K., C.P. Rajendran, M. Thakkar, and M. P. Tuttle, 2001, The 2001 Kutch (Bhuj) earthquake: coseismic surface features and their significance, *Curr. Sci., 80*, 1397–1405.

Rothaus, R.M., E. Reinhardt, and J. Noller, 2004, Regional consideration of coastline change, tsunami damage and recovery along the southern coast of the Bay of Izmit, *Nat. Hazards, 31*, 233–252.

Wang, C.-Y., A. Wong, D.S. Dreger and M. Manga, 2006, Liquefaction limit during earthquakes and underground explosions: Implications on ground-motion attenuation, *Bull. Seism. Soc. Am., 96*, 355–363.

Yu, M.-S., B.-C. Shieh, and Y. T. Chung, 2000, Liquefaction induced by Chi-Chi earthquake on reclaimed land in central Taiwan, *Sino-Geotech, 77*, 39–50.

E.5 Triggered Mud Volcanoes

Earthquake	Magnitude	Epicenter distance (km)	References
Tokachi-Oki, 16 May 1968	8.2	186	Chigara and Tanaka (1997)
Tokachi-Oki, 4 Mar 1952	8.6	58	Chigara and Tanaka (1997)
Japan, 21 Mar 1982	6.7	23	Chigara and Tanaka (1997)
Kushiro-Oki, 15 Jan 1993	7.6	153	Chigara and Tanaka (1997)
Hachinohe, 28 Dec 1994	7.8	226	Chigara and Tanaka (1997)
Tokachi-Oki, 25 Dec 2003	8.3	145	Manga and Brodsky (2006)
Sumatra, 26 Dec 2004	9.1	900	Manga and Brodsky (2006)
Azerbaijan, 28 Jan 1872	5.7	24, 40	Mellors et al. (2007)
Azerbaijan, 13 Feb 1902	6.9	45, 51	Mellors et al. (2007)
Makran, 28 Nov 1945	8.1	41, 155, 189	Delisle (2005)
Baluchistan, 30 May 1935	7.7	61	Snead (1964)
Caspian, 8 Jul 1895	8.2	141	Mellors et al. (2007)
Romania, 4 Mar 1977	7.2	92	Mellors et al. (2007)
Azerbaijan, 24 Sept 1848	4.6	15	Mellors et al. (2007)
Mongolia, 4 Dec 1957	8.3	75	Rukavickova and Hanzl (2008)
Mongolia, 15 Jun 2006	5.8	90	Rukavickova and Hanzl (2008)
Pakistan, 26 Jan 2001	7.7	482	Deville (submitted for publication)
Regnano, Italy, 11 Oct 1915	5.0	21	Martinelli et al. (1989) in Bonini (2009)
Italy, 4 Sept 1895	5.0	4	Trabucco (1895) in Bonini (2009)
Montegibbio, Italy, 5 Apr 1781	5.8	87	De Brignoli di Brunhoff (1836) in Bonini (2009)
Montegibbio, Italy, 91 BCE	5.7	15, 16	Guidoboni (1989) in Bonini (2009)
Paterno, 13 Dec 1990	5.7	39	D'Alessandro et al. (1996) in Bonini (2009)
Paterno, 4 Oct 1978	5.2	34	Silvestri (1978) in Bonini (2009)
Italy, 5 Mar 1828	5.9	56	La Via (1828) in Bonini (2009)

References

Bonini, M., 2009, Mud volcano eruptions and earthquakes in the Apennines, Italy, *Tectonophysics, 474*, 723–735.
Chigira, M. and K. Tanaka, 1997, Structural features and the history of mud volcanoes in Southern Hokkaido, Northern Japan, *J. Geol. Soc.Japan, 103*, 781–791.
Delisle, G., 2005, Mud volcanoes of Pakistan – An overview. In: Martinelli and Panahi (eds.), *Mud Volcanoes, Geodynamics and Seismicity*, pp. 159–169.
Deville, E., 2008, Mud volcano systems, *New Research on Volcanoes, Eruptions and Modeling*, Nova publishers, in press, ISBN 978-1-60692-916-2.
Manga, M., and E. Brodsky, 2006, Seismic triggering of eruptions in the far field: Volcanoes and geysers, *Ann. Rev. Earth Planet. Sci., 34*, 263–291.
Mellors, R., D. Kilb, A. Aliyev, A. Gasanov, and G.Yetirmishli, 2007, Correlations between earthquakes and large mud volcano eruptions, *J. Geophys. Res., 112*, B04304.
Rukavickova, L., and P. Hanzl, 2008, Mud volcanoes in the Khar Angalantyn Nuruu, NW Gobi Altay, Mongolia as manifestation of recent seismic activity, *J. Geosoc., 53*, 181–191.
Snead, R.E., 1964, Mud volcanoes of Baluchistann, West Pakistan, *Geograph. Rev., 54*, 546–560.

E.6 Triggered Earthquakes

Earthquake	M	Date	Epicenter distance (km)	References
Denali	7.9	2002/11/3	3457	Brodsky and Prejean (2005)
Landers	7.3	1992/6/28	438	Brodsky and Prejean (2005)
Hector Mine	7.1	1999/10/16	397	Brodsky and Prejean (2005)
Mendocino	7	1994/9/1	683	Brodsky and Prejean (2005)
Northridge	6.7	1994/1/17	381	Brodsky and Prejean (2005)
San Simeon	6.5	2003/12/22	302	Brodsky and Prejean (2005)
Eurake Valley	6.1	1993/5/17	102	Brodsky and Prejean (2005)
Parkfield	6	2004/9/28	243	Brodsky and Prejean (2005)
Double Spring Flat	5.9	1994/9/12	149	Brodsky and Prejean (2005)
Northridge aftershock	5.8	1994/1/17	379	Brodsky and Prejean (2005)
Ridgecrest	5.5	1995/9/20	355	Brodsky and Prejean (2005)
Eureka aftershock	5	1993/5/18	114	Brodsky and Prejean (2005)
Tokachi-oki	8.3	2003/9/25	1340	Miyazawa and Mori (2005)

(continued)

Earthquake	M	Date	Epicenter distance (km)	References
Tokachi-oki	8.3	2003/9/25	1110	Miyazawa and Mori (2005)
Tokachi-oki	8.3	2003/9/25	920	Miyazawa and Mori (2005)
SW Siberia	7.3	2003/9/27	4150	Miyazawa and Mori (2005)

References

Brodsky, E.E., and S.G. Prejean, New constraints on mechanisms of remotely triggered seismicity at Long Valley Caldera, *J. Geophys. Res., 110*, B04302, doi:10.1029/2004JB003211.

Miyazawa, M., and J. Mori, 2005, Detection of triggered deep low-frequency events from the 2003 Tokachi-oki earthquake, *Geophys. Res. Lett., 32*, L10307, doi:10.1029/2005GL022539.

Index

The manufacturer's authorised representative in the EU is Springer Nature Customer Service Centre GmbH, Europaplatz 3, 69115 Heidelberg, Germany. If you have any concerns regarding our products, please contact ProductSafety@springernature.com

Printed and bound by CPI Group (UK) Ltd, Croydon, CR0 4YY
30/09/2026
02225643-0005